T0202186

Disembodied Brains

Disembodied Brains

Understanding Our Intuitions on Neuro-Chimeras and Human-Brain Organoids

JOHN H. EVANS

OXFORD
UNIVERSITY PRESS

OXFORD
UNIVERSITY PRESS

Oxford University Press is a department of the University of Oxford. It furthers
the University's objective of excellence in research, scholarship, and education
by publishing worldwide. Oxford is a registered trade mark of Oxford University
Press in the UK and certain other countries.

Published in the United States of America by Oxford University Press
198 Madison Avenue, New York, NY 10016, United States of America.

Library of Congress Cataloging-in-Publication Data
Names: Evans, John Hyde, 1965– author.
Title: Disembodied brains : understanding our intuitions on
neuro-chimeras and human-brain organoids / John H. Evans.
Description: New York, NY : Oxford University Press, [2024] |
Includes bibliographical references and index.
Identifiers: LCCN 2023033836 (print) | LCCN 2023033837 (ebook) |
ISBN 9780197750704 (hardback) | ISBN 9780197750728 (epub) |
ISBN 9780197750711 (ebook) | ISBN 9780197750735 (ebook other)
Subjects: MESH: Neurosciences—ethics | Brain | Chimeras | Organoids | Bioethical Issues
Classification: LCC QP376 (print) | LCC QP376 (ebook) |
NLM WL 21 | DDC 612.8/2—dc23/eng/20230830
LC record available at https://lccn.loc.gov/2023033836
LC ebook record available at https://lccn.loc.gov/2023033837

DOI: 10.1093/oso/9780197750704.001.0001

Printed by Integrated Books International, United States of America

Contents

Acknowledgments

In my previous contributions to the field of public bioethics, I have examined both the expert debate and what the general public thinks about a given topic. I have largely focused on what is now called human gene editing, and a large part of that debate is, as one would expect, focused on what a human being actually is. Over the years I was always aware of an adjacent debate about human-animal hybrids and chimeras but never read deeply into the debate. In 2020 I was asked to be a member of the Committee on Ethical, Legal, and Regulatory Issues Associated with Neural Chimeras and Organoids, sponsored by the National Academies of Sciences, Engineering, and Medicine. Unfortunately, since these reports have deadlines, I was unable to spend many months thinking about all the implications of the technologies considered, and certainly lacked the time to create my own empirical research project. Now I have finally had the time and distance to conduct an empirical examination.

I greatly appreciate my fellow committee members, with whom I have had such productive conversations and respectful disagreements: Bernie Lo, Josh Sanes, Paola Arlotta, Alta Charo, Rusty Gage, Hank Greely, Pat King, Bill Newsome, Sally Temple, Larry Zipursky, and Anne-Marie Mazza. If only we could have met live instead of on Zoom.

Thanks also to Ross Graham for general research assistance, to Mike McCullough and Seth Hill for advice on the survey, and to Richard Pitt for helping to interpret some survey findings.

Additional thanks to reviewers from the Press for suggesting improvements in the original text, and to Peter Ohlin for shepherding this process. And, as always, thanks to Ronnee Schreiber for continuing to put up with me and my strange academic interests for more than three decades now.

1

The Science and Public Ethics of Neuro-Chimeric Animals and Human-Brain Organoids

In one of the most successful films of 1968, moviegoers saw Charlton Heston star as an astronaut who travels to a strange planet,[1] where he encounters a species of intelligent talking apes who dominate the mute and primitive humans who share the planet. Surely this movie was popular at least partly because of the acting and the story, but *Planet of the Apes* was also, in its time, the latest example of our society's fascination with animals having human qualities, which I will call human-animal *chimeras*. This fascination has been with us for thousands of years.

The term *chimera* comes from ancient mythology and refers to the mixing of any two or more diverse animals. The original mythological chimera was a mix of lion, bird, and snake. But the category of chimera in mythology also includes human-animal mixes such as sphinxes (half human, half lion), centaurs (half human, half horse), satyrs (half human, half goat), minotaurs (half human, half bull), and sirens (half human, half sea monster).[2]

Contemporary culture has produced an enormous number of an even more specific type of chimera—that is, an animal who thinks and acts like a human, like those on *Planet of the Apes*. If the movie's apes had had human stomachs nobody would have cared, but *neuro*-chimeras (NCs) mix in the human brain, which is considered by many to be what makes us human.

Disembodied Brains. John H. Evans, Oxford University Press. © Oxford University Press 2024.
DOI: 10.1093/oso/9780197750704.003.0001

In popular culture, NCs are not necessarily evil. An older generation will remember the 1960s TV show about the talking horse Mr. Ed, whose speech could only be heard by his human friend Wilbur. And cartoons are filled with animals that act like humans: Mickey Mouse, Daffy Duck, Bugs Bunny, Stuart Little, and many more.

The fascination with the Planet of the Apes is just the latest installment of a long story. If that movie had concerned standard apes, it would obviously not have had the same sensational impact. Chimeras, by contrast, have cultural power and provoke fascination. In Robert Klitzman's summary: "the long history of such images across multiple, disparate cultures suggests how humans have long seen the boundaries between species as fluid, but in differing ways, as either good or bad—as gods or monsters—in both cases as special and possessing power."[3]

I am usually loath to use the term "impossible" when it comes to scientific developments, but an animal body that thinks and speaks exactly like a human, such as Mr. Ed or Bugs Bunny, will probably always be impossible. These fictional entities would need a human brain, and we are only one tiny step down the path to Mr. Ed.

In this book one of my foci will be human-animal NCs, which are animals with some human brain tissue. While we do not know how Mr. Ed came into being, presumably he is an NC, because he thinks and speaks like a human, albeit trapped in a horse body that limits his human-like abilities. While Mr. Ed was fascinating and entertaining, an actual Mr. Ed would be unsettling and intensely controversial.

Let me turn to another trope from popular culture: the brain in the vat. The poster for the 1962 B-movie *The Brain That Wouldn't Die* shows a woman's head and a brain in a large vat, with wires running between the head and the vessel, and the words "Alive . . . Without a body . . . Fed by an unspeakable horror from hell!"[4] In the more contemporary movie *The Matrix,* computers have enslaved human bodies, having plugged their brains into a simulation where these brains all think that they are having normal human lives even as they actually remain immobile in a vat.[5] While the brain in the vat, unlike the chimeras, has not been a source of centuries-long fascination, it reflects a host of longtime cultural distinctions, such as the mind-body problem, the limitations of human perception, and the horror of perceiving social life without being able to engage in it.

There are reasons why the talking apes and the brain in the vat sell movie tickets. They touch on the deepest questions of human existence, and their stories allow us to work out these deep questions without completing a PhD in the philosophy of mind. But they also permit the viewer to walk out of the theater without a sense of looming disaster, because they are just stories, right?

Is the Fictional Becoming Real?

What happens when human-animal NCs and disembodied brains in a vat stop being quite so fictional? We have been on the path of *nonneural* chimeras for centuries, depending on which historical claims you believe and what you include in that category. For example, taking a part of an animal and putting it in a human, or *xenotransplantation*, results in a type of human-animal mix. It has been claimed that a bone from a dog was put into a human in 1501, producing the first known (non-neural) human-animal chimera. In 1667 a man was infused with lamb's blood. In the following centuries doctors would continue to attempt similar feats, without much success.[6]

To turn to the modern era, in the early 1960s various Americans experimentally received organs from baboons and rhesus monkeys, though even the most successful transplants resulted in patients surviving only a few months. In 1984 a highly publicized transplant of a baboon heart into a baby named Fae failed; according to one historical summary, "Most of the hopes put on xenotransplantation died with Baby Fae."[7]

Nevertheless, the invention of better immunosuppressive drugs, and the creation of genetically modified animals whose organs are more compatible with humans, led to further attempts. In 1992 the first transgenic pig, with genetically engineered DNA, was born, and the success would lead to the widespread use of pig organs in humans.[8] Today, implanting heart valves from both pigs and cows is standard medical practice, and scientists continue working on genetically modifying pigs to make their organs more compatible with humans.[9] Other nonneural chimeras include rodents whose immune systems have been altered to more closely resemble the human immune system.[10]

Since blood, livers, and the immune system do not raise foundational questions about humanity, I will put these human-animal combinations aside for the rest of this book. Our concern here will instead be the brain. In modern times it is the brain that is seen as the core of what makes us different from other life-forms and the seat of the self. With brain death as the legal standard for being dead, the brain is also how we decide which human bodies live or die.

We should naturally ask why scientists are doing this research. Over the past few hundred years, the scientific motive has largely been to save lives by replacing dysfunctional human parts with functional animal parts. But the new research on NCs and brains in a vat has a slightly different motivation, which is to create experimental animal models for studying human neurological disease.

Since we cannot experiment on actual live human brains because they are residing in live humans, alternative experimental models are needed. The brains of lab mice are quite different from those of humans, so these experiments do not tell us as much about human neurological disease as we would like. There is an inherent tension in employing such human-brain substitutes: the more they resemble human brains, the more useful they are for human research, but also the more ethical issues are raised.

Development of Human-Brain Organoids

To better understand these new technologies, let us start with the brain in the vat: that is, a *human brain organoid*, or HBO. An HBO is a three-dimensional aggregation of human neural cells grown from pluripotent stem cells. A stem cell is a cell that makes other cells, and a pluripotent stem cell can be used to make a range of specific cell types.

Scientists have successfully grown cells that make up one small part of a brain (not an entire brain) to produce a clump of tissue about 4 mm across that can be kept alive in a dish. And researchers can now take HBOs of different parts of the brain and coax them into fusing to become "assembloids." Importantly, these entities are kept in a petri dish except when they are placed in an animal brain (see below). The concern of ethicists and scientists about HBOs in a dish is that they could achieve a modicum of consciousness (about which *much* more below). Philosophers would say that an entity cannot have higher-level consciousness without experiences; thus, with no way of interacting with the world, HBOs will remain lumps of tissue.[11] However, HBOs representing tissue from different parts of the human brain have recently been connected, and, in a 2017 article, scientists reporting an experiment on HBOs with retinal cells found that a burst of electrical activity followed from shining light on the HBOs.[12] Scientists have also had success in growing optic components of the brain.[13] In 2019, researchers found that the electroencephalogram activity of an HBO resembles the activity seen in 25- to 39-week-old premature infants.[14] This all has researchers wondering whether HBOs might be more capable of acquiring abilities than previously assumed.

One of the primary HBO researchers has created a small spider-like robot whose movement is stimulated by the electrical activity of the HBO, and he recently announced that he is working on "closing the loop" so that the perceptions of the robot will be fed back to the HBO. Another scientist, working in conjunction with

Microsoft Corporation to improve machine learning, has said he is pursuing the problem of connecting the "wetware" of the HBO to the "hardware" of a computer to make a more efficient computer.[15] Perhaps it is not too far-fetched to predict that within ten years an HBO in a dish could be having rudimentary experiences.

This research has received a lot of media attention, partly because there had not previously been any research like it. Of course, scientists have been growing human tissue from stem cells for quite some time, and the HBO techniques have built on that research. But culturally the brain is something altogether different from a liver or even a heart. The headlines suggest the potential explosiveness of the continuing experimental progress. A headline from *Scientific American*, "Lab-Grown 'Mini-Brains' Can Now Mimic the Neural Activity of a Preterm Infant,"[16] evokes the image of a baby trapped in a petri dish. So we can see why attention has been focused on HBOs.

Development of Human-Animal Neuro-Chimeras

Scientists' ability to create human-animal NCs has recently increased markedly. The first technology that implied the possibility of a NC was "human-to-animal embryonic chimeras," produced by putting human neural stem cells into the brain of a developing monkey and putting human genetic material into rabbit eggs.[17] A 2005 report of the National Research Council and the Institute of Medicine urged restraint on developing these and similar NCs, stating that "perhaps no organ that could be exposed to [human embryonic stem cells] raises more sensitive questions than the animal brain, whose biochemistry or architecture might be affected by the presence of human cells."[18]

Scientists have since implanted human glial cells into mice, who then—not to get too ahead of myself—performed better on learning tests.[19] In 2018 HBOs were first implanted into the brains of adult mice to create a NC, combining the two technologies under examination in this book.[20] In an experiment reported in the fall of 2022, scientists made striking progress in both HBO and NC research by implanting HBOs in the brains of two- to three-day-old rats.[21] While HBOs in a dish are not vascularized, and thus the middle of the HBO cannot receive oxygen and nutrients, limiting its growth, an HBO in a rat becomes part of the rat's circulatory system and thus can grow larger. In the 2022 experiment, each cell in the organoid grew six times longer than it would have grown in a dish, and the cells became about as active as neurons in human brains.[22] This seemed to solve the problem that HBOs in a dish would always be small and limited, in that the rat could become the vat that allowed the HBO to grow to a more human level. Another problem with the HBO in the dish is that brain neurons develop with experiences, and it is hard to have experiences in a dish—but as the rat has experiences, its HBO becomes more developed and human-like.

What the experiment showed about brain integration, and thus about the development of NCs, is even more morally relevant than what it showed about HBOs. The experiment suggested a true mix of human and rat brain, instead of an HBO simply residing, unintegrated, in the vat of the rat brain. When implanted, the HBOs spontaneously wired themselves with even distant parts of the rat brain. The human component also reacted to stimulation of the rat's whiskers, and even directed rat behavior, with the organoids "apparently sending messages to the reward-seeking regions of the rats' brains." Unlike in the earlier experiments, however, these rats did not become any smarter.[23]

The *New York Times* report of the research quotes a scientist assuring us that the rats are still rats and are not human. But, the writer notes, "that might not hold true if scientists were to put human organoids in a close relative of humans like a monkey or a chimpanzee." The author of the study pointed out to the reporter that the similarity between primates and humans might allow the organoids to grow larger and take on a larger role in the animal's mental processes, but added that "it's not something that we would do, or would encourage doing."[24] But while this particular researcher would not do such research, we can be confident that the ability to make a monkey brain more like a human one will attract other researchers trying to relieve the suffering from human brain diseases.

A second method of creating an NC occurs much earlier in the life of the animal: putting human embryonic stem cells into animal blastocysts (advanced-stage embryos), resulting in an animal with some human cells. While this has not yet been done with brain cells, recent research has made such experiments more possible.[25] For example, by means of a technique called blastocyst complementation, a pig blastocyst could be modified to open a developmental niche that would otherwise have developed into a pig pancreas. Human stem cells could be placed into that niche, and the resulting pancreas in the pig would be like a human pancreas.[26]

In 2018 researchers first conducted *neural* blastocyst complementation, taking the developmental niche that would have developed a mouse brain and replacing it with stem cells from another type of mouse.[27] This opens the door for future experiments in which a mouse could develop with human brain tissue.

Scientists have begun research on nonhuman primate NCs using monkeys. This research is motivated by the familiar problem with using rodents for human research, which is that rodents diverged from humans in evolution so many millions of years ago that we humans are really not like lab rats, particularly in our brains.[28] This frustration is exemplified by the common quip by scientists that it is a great time to be a mouse with Alzheimer's, since so many drugs have been shown to be efficacious on Alzheimer's in mice even though they do not end up working in humans. By contrast, nonhuman primates have not only greater genetic similarity to humans but also more similar skull size and gestation time than other mammals, and their bodily form also means that they could have more humanlike experiences.

In 2021 scientists in China announced that they had successfully added human cells to monkey blastocysts and that these human cells had survived far into development.[29] Though these were not specifically neural cells, this research produced the proof of concept that it is possible to create monkey-human chimeras by merging cells in a blastocyst.

A third method of creating an NC is engineering a human-animal transgenic hybrid through genetic modification. For decades researchers have been producing "humanized mice," mice genetically modified to have human qualities.[30] Even NCs have been created, such as mice genetically engineered to have Alzheimer's disease and monkeys with human autism.[31] In 2019 scientists put a human gene that is thought to affect intelligence into a monkey genome, and the altered monkeys were reported to be smarter than comparable unaltered monkeys.[32] Scientists are pressing forward with developing transgenic hybrid nonhuman primate models for human neurological disease.[33]

These experiments have been deeply controversial. Again, a few headlines will make the point. "Chinese Scientists Have Put Human Brain Genes in Monkeys—and Yes, They May Be Smarter" was the title of an article in one science magazine.[34] "The Smart Mouse with the Half-Human Brain" was a headline in The New Scientist.[35]

To make my point about how controversial these technologies will be, I would request that you describe HBOs and NCs to someone in your home and ask what they think about them. I predict you will hear expressions of both shock and disgust that such research exists (of which more below). We can expect this research to become increasingly controversial as the public finds out more about it.

From this point forward I will not focus on the details of how NCs and HBOs are created, because the exact method lacks moral valence. The central question at this point is: Should this research continue? The answer will depend upon our ethics.

What Should We Do?

Whose ethics get to shape public policy regarding biomedical technology? Before roughly the 1960s in the United States, the answer would have been simple: scientists' ethics. But since the 1960s the answer has changed, and it is now, at least officially: the public's ethics. The reader will then ask how—in a morally and religiously pluralistic society, where even summarizing the ethics of all the citizens would be extremely difficult—it is possible to allow the public's ethics to decide such policy, even assuming that the regular public had the time to stay up to date on these technologies.

In order to determine the public's ethical stance, the tradition has been to use a group of experts who ostensibly represent the public's views to mediate between the public and policy makers. At present, elected officials and the officials they appoint ultimately make policy decisions on these issues; however, critically, the ethical choices considered to be legitimate are filtered, discarded, or promoted by the experts before policy makers rule on an issue. This filtering of possible legitimate ethical stances very powerfully influences what policy is made, and ultimately what technologies are pursued. In the United States, this structuring occurs in "public-policy bioethical debate"—that is, the debate about the ethics of the technology engaged in among elite experts, essentially all of whom have doctoral degrees, making recommendations about what an ethical policy should be. Such debate occurs largely in specially created commissions, conferences, and academic publications.[36]

When the public-policy bioethical debate began in the 1960s and 1970s, a broader range of elites was involved. Of those previously involved, those currently absent include, most notably, the pluralistic theologians, who spoke a secular language in public and were focused on the deeper implications of technologies. Today the debate is instead dominated by scientists and bioethicists.

The profession of bioethics is like philosophy in that it makes normative claims, but it is distinct from philosophy in the types of arguments it makes.[37] A typical debate occurs when a new and controversial technology arrives on the scene, and a group of scientists and bioethicists (with scattered others) is convened to ascertain what the ethical policy should be. They start with the public-policy bioethical debate, which is published in academic journals. They write reports, which scientific agencies use to set policy and practicing scientists generally follow voluntarily. On the topics of HBOs and NCs, there have been National Institutes of Health committees, committees of scientific associations, and committees of the National Academies of Sciences, Engineering, and Medicine.[38]

The scientists and the bioethicists generally agree on the values and goals that should drive the ethics of technologies. To understand the ethical policies about HBOs and NCs that have already been recommended—and to some extent enacted—we must understand the values advanced by both groups in these debates.

Before making a statement about these two groups, I need to say that this book is based in sociology, which means that it will be making generalizations. When a sociologist says that "people in Alabama are more conservative than people in California," he or she simply means that *on average* Alabamans are more conservative than Californians. Similarly, when a sociologist refers to the opinion of "scientists," he or she means the average or dominant scientists. No group is monolithic.

That said, the bioethicists and scientists advocate for the use of only a few purportedly universal goals or values: autonomy (freedom), beneficence (doing good), non-maleficence (avoiding harm), and justice (equal treatment).[39] To oversimplify: If a technology does not harm but helps, if people are free to use or not use it, and if it is equally available to all—then it is morally acceptable.

In contrast, scholars who do not participate in the public-policy bioethical debate are often using other values. The bioethical debate about HBOs and NCs that is currently engaged in for guiding policy is focused only on the value of non-maleficence. Beneficence is assumed to be advanced in medical research, while justice and autonomy are only relevant for persons as opposed to animals. The question in the present debate is: Is HBO technology harmful for the HBO, and is the inclusion of human tissue in an animal harmful for the NC?

I do not think bioethicists and scientists are being nefarious in limiting the values under consideration. In many cases, this limited set of values comprises the only ones that these experts truly believe in. In other cases, the experts believe that, in a liberal democratic society, only nearly consensual imperatives such as "Don't harm people" can be imposed. My point is that these are actually not the only values important to the public.

Public Input to the Public-Policy Bioethical Debate

From the end of the twentieth century, the academic experts in the public-policy bioethical debate have recognized that they have a legitimacy problem. On what grounds should this unusual group of citizens—all of whom have PhDs—get to structure government policy? They are not elected, and they are not representative of the general public. There has been a slowly growing recognition that, since the debate provides input to public policy, it should therefore promote not simply the values of scientists and bioethicists but more broadly the values of the public. This is not simply for general democratic reasons but also because most of this research is paid for by the public.

In recognition of this problem, bioethicists and scientists now regularly call for the input of the public's ethics into policy deliberation through public-engagement exercises. Each of the aforementioned bioethics reports has a passage at the end calling for such engagement. However, such exercises do not actually influence policy, since they lack "teeth." According to one interpretation from prominent science-communication scholars, they simply give the appearance that the public's values are being followed.[40] Moreover, despite the ubiquitous call for such public-engagement exercises, they never seem to happen, leaving the expert ethics of the bioethicists and scientists to continue setting the agenda.

I have elsewhere proposed a solution to this problem, which is that the values of the public should serve as the raw material from which bioethicists can derive more detailed ethical recommendations.[41] The values of the public on a particular issue would be described by social science. It is in that spirit that this book will proceed, contrasting the public's views with those of the bioethicists and scientists who dominate the public-policy bioethical debate.

Defining the Human

To understand the difference between the public's ethics regarding HBOs and NCs and those of the bioethicists and scientists, we have to briefly delve into an admittedly abstract concept that is almost never explicitly discussed in these debates. This concept is the definition of the human, and we have to explicitly raise it because the public-policy bioethical debate presumes that the definition used by bioethicists and scientists is an undebatable fact. HBOs and NCs are controversial because of their connection to actual humans, so it should not surprise us that one's perspective on these technologies is dependent on what a human actually is. The "avoiding harm" value that is most prominent in the debate applies centrally to humans (and, to a lesser extent, to animals with more human-like capacities).

Among academics in the Western intellectual tradition, there are three prominent definitions of the human, also called "anthropologies": the theological, the philosophical, and the biological.[42] These definitions can be based on the physical form of the human, the moral status of the physical form, or both, and they typically contrast humans with other animals.

Theological Anthropology

Among the various theological anthropologies, I focus on the Christian anthropology, because Christian beliefs, unlike those of other religions, have historically influenced the public culture and law in the United States, even for nonreligious people. (Due to Christianity's emergence from Judaism, the Christian anthropology is very similar to the Jewish version.) Christianity is also by far the largest religious tradition in the United States, and, as I will describe below, it is likely the source of the greatest variation in the public's views. More pragmatically, when I later turn to obtaining the views of the public, it will be evident that non-Christian U.S. religious groups are too small to make claims about using the methods employed in this book.

For the Christian theological anthropology, the central idea is that humans are defined by being made in the image of God. The account of human origins in the book of Genesis (1:26-28) reads: "Then God said, 'Let us make man in our image, after our likeness; and let them have dominion over the fish of the sea, and over the birds of the air, and over the cattle, and over all the earth, and over every creeping thing that creeps upon the earth.' So God created man in his own image, in the image of God he created him; male and female he created them" (Revised Standard Version). Note that this critical biblical passage not only gives the definition of a human but also more broadly includes a description of the human relationship to animals and nature, which I will discuss below.

There are undoubtedly many thousands of pages of academic theological texts devoted to what exactly the image of God is, but it is *not* the case that a Christian theologian identifies physical humans by using a checklist of attributes of God. Rather, to identify a *physical* "human," the Christian anthropology relies upon the pre-DNA biological definition of a human, in which humans are those born from other humans (of which more below). In the very traditional version, humans are the descendants of Adam and Eve, the first two humans.

The first component of this anthropology is that humans are not limited to physical form, but also have a "soul." When creating each human, one by one, God gives them each a soul. This soul is, for the academic theologians, essentially communicative—it is how a human relates to other humans and to the divine.[43] Traditionally, this soul can separate from the body and will travel to heaven (or hell) after death. Thus, humans are not just material; they are also spiritual beings. This is the one anthropology that is not materialist, maintaining that there are true features of the world beyond atoms, electricity, and other phenomena that can ultimately be explained with physics.

This anthropology defines not only physical form but moral status as well. First, physical humans are not animals, which were created separately by God, with humans as their stewards. This anthropology has *traditionally* emphasized that humans are special compared to other animals, as it is only humans who are made in God's image and only humans who communicate with God. In this view, each human is individually ensouled by God, often described as occurring at fertilization.

Of course, perspectives on the moral status of humans compared to animals are not uniform. Protestants typically make a categorical distinction between human and nonhuman animals. One evangelical scholar writes that "Evangelicalism is influenced by long held Christian and social traditions which emphasize a categorical distinction between humans and animals: *imago dei* that sets humanity apart from the rest of creation; whereby humans are given dispensation to use animals under God." Many Conservative Jewish scholars agree, with one scholar writing that there is "admiration for each of the various species God has made, with no hierarchy of value among them—except for humans, who are, according to Genesis (1:27; 5:1; 9:6), uniquely created in the image of God. This special status permits us to use animals for our purposes."[44]

Catholicism is a bit more ambiguous, as exemplified by St. Francis of Assisi and the blessing of the animals that is widely conducted in Catholic churches today. Many contemporary Catholic theologians want to at least lessen the distinction between human and nonhuman animals by saying that both bear the image of God. In this account, humans and animals are still distinct, but animals are not solely for human use. Nonhuman animals were created by God (on the same day, in the Genesis account), and each animal has its own telos that must be respected. As two theologians write: "God created animals "good," *period*, to flourish in their own right as the good kinds of things that they are."[45] We then cannot think of animals as tools for human use.

The second moral status component of this anthropology is between humans. To return to the Christian consensus: since each human was created individually by God, and communicates with God, all humans have equal value—an ethic that adherents of this anthropology have obviously not always upheld. It then emphasizes that being made in the Image—and thus being human—is not dependent upon any capacities of the human. One classic explanation is that humans have "alien dignity," because their dignity does not come from any capacity they possess but rather from God.[46] This explains the tendency for Christians to less readily accept that biologically human entities that lack certain capacities, like fetuses or the comatose, can be destroyed.

The theological anthropology is then both biological (a definition of the physical difference between humans and animals) and moral: all humans are to be treated as equals without regard to capacities and are in a superior position to all animals. This definition of the human, while more abstract than the other anthropologies, has had enormous influence in the West, such as being the origin of secular human rights,[47] and is widely held by Americans.[48]

Philosophical Anthropology

The philosophical anthropology ignores the physical form of the entity and the distinction between humans and animals while focusing on moral status. It does not talk about the human, but about persons. According to Peter Singer, its best-known proponent, "the concept of a person is distinct from that of a member of the species Homo sapiens, and . . . it is personhood, not species membership, that is most significant in determining when it is wrong to end a life."[49]

Personhood is the moral status an entity reaches if it has enough valued capacities. There is rough agreement about at least the most basic capacities required for personhood. In one list of 17 capacities, highlights include "possesses consciousness," "can experience pleasure and pain," "has thoughts," "has preferences," "is capable of rational deliberation," and "has a sense of time."[50] For Peter Singer, personhood requires being, "at minimum, a being with some level of self-awareness."[51]

Using this anthropology, human fetuses and those in permanent vegetative states are not persons because they lack important capacities, and there is thus no prohibition on ending their lives. Chimpanzees and some other nonhuman animals, on the other hand, are often thought to have the capacities required for personhood and therefore should be treated accordingly.

The philosophical anthropology is interwoven with a wide range of bioethical policies. For example, a living will contains a list of capacities in whose absence the person does not want to be kept alive. All regulations concerning the treatment of animals are loosely based on this anthropology, within which, for example, you can do anything you want to ants, which have no valued capacities, but cannot experiment on chimpanzees, who have many of the valued capacities. The critical question is how many valued capacities a life-form has.

Biological Anthropology

The biological anthropology is resolutely materialist, with no conception of soul or spirit. It has classically been concerned only with physical form—with distinguishing the human species from other animals. Ernst Mayr's classic "biological species concept" holds that a species is defined by reproductive isolation: if two entities cannot breed, they are from separate species. If we start with those who are consensually considered humans, then any entity that can breed with those people would also be human. Since chimps cannot breed with humans, they are not human. A related classic definition, the "evolutionary species concept," emphasizes continuity: "a species is a single lineage of ancestral descendant populations of organisms which maintains its identity from other such lineages, and which has its own evolutionary tendencies and historical fate."[52] That is, humans are those born from humans, which is akin to the biological components of the theological anthropology.

However, these anthropologies are no longer dominant among biologists. As Jason Scott Robert and Françoise Baylis write, "As against what was once commonly presumed, there would appear to be no such thing as fixed species identities" for biologists. Post-DNA biology has emphasized the evolutionary idea that humans are not distinct but are on the same continuum with other animals. One critic has assembled statements from influential biologists to this effect: "Humans are more like worms than we ever imagined," "The worm represents a very simple human," "In essence, we are nothing but a big fly," "We share 99 percent of our genes with mice, and we even have the genes that could make a tail," "We humans appear as only slightly remodeled chimpanzee-like apes," and so on.[53]

While in principle the biological anthropology makes no moral status claims, and it is part of the ideology of science to be morally neutral, the species-continuum view of physical form coincides nearly perfectly with the philosophical anthropology. The ranking of capacities that are critical to the philosophical anthropology map onto evolutionary differences, and therefore the evolutionary continuum (microbes to insects to mammals to humans) very closely coincides with the valued-capacities continuum. It is the biological forms of life most evolutionarily distant from humans that have the least-valued capacities, and those evolutionarily closest (chimpanzees) that have the most-valued capacities.

Becoming Human through Capacities?

Armed with these definitions of the human, we can now properly describe the public-policy bioethical debate about HBOs and NCs that is dominated by bioethicists and scientists. This debate largely assumes the philosophical anthropology and is concerned with whether HBOs or NCs could obtain the requisite capacities to "become human." Of course, no one in the mainstream debate thinks that an NC mouse or an HBO could become a physical or biological human; however, the biological distinction between species is not relevant for those using the philosophical anthropology. Since developing enough valued capacities to qualify for full personhood—to be morally treated like other humans—is unlikely, another question becomes whether the entity has enough valued traits to be treated like we treat more advanced animals.

The base capacity discussed in this debate is "consciousness" or "sentience," as well as the related concept of "cognitive ability."[54] Henry Greely is the most prominent bioethicist examining NCs and HBOs, and his work reflects the anthropological assumptions among bioethicists in the bioethical debate. Greely is concerned with the "humanization" of NCs, and he defines humanization as having "some human cognitive abilities"—a topic he marks as the most significant part of the debate.[55] Greely similarly writes of HBOs: "If it looks like a human brain and acts like a human brain, at what point do we have to treat it like a human brain—or a human being?"[56] In other anthropologies, by contrast (as I will discuss below), a disembodied brain would never be considered a human being.

A section of another paper from Greely is titled "Conferring Humanity on Mice," which suggests that mice could obtain human moral status. Greely and his colleagues set the standard for whether you could "transform a mouse into a man" by whether it was "a creature with some aspects of human consciousness or some distinctively human cognitive abilities."[57] This anthropology is not only used by Greely. H. Isaac Chen and colleagues write that "One of the primary concerns centers on the possibility that animals transplanted with human brain organoids would become more 'human.'" Their ethical concerns about HBOs are due to HBOs' being "increasingly similar to the human brain, the source of the higher-order cognitive capacities that are most often equated with being human."[58]

The assumption that consciousness is the primary moral consideration is also evident in the work of the bioethics commissions dedicated to this topic. The National Academies Committee assigned to study HBOs and NCs was given a list of questions to answer. The first question on the list was "How would researchers define or identify enhanced or human awareness in a chimeric animal?" The second was "Do research animals with enhanced capabilities require different treatment compared to typical animal models?" The fifth was "How large or complex would the ex vivo brain organoids need to be to attain enhanced or human awareness?" The ethical concerns of the public, which I describe below, are not mentioned as topics to be considered.[59]

Media stories also tend to take this perspective and use the philosophical anthropology to compare the capacities of HBOs and NCs with those of born humans, partly because they rely on bioethicists as their experts. For example, the *New York Times* article reporting on the discovery that brain waves in HBOs resemble those of premature babies is centrally concerned with whether this means that the two entities are effectively the same. In the article, the scientist whose lab made the discovery stated his concern that HBOs could become "conscious," and neuroscientist Christof Koch worried about a "brain that is capable of sentience and of feeling pain, agony and distress."[60]

Similarly, in a news article describing the NC mice that resulted from the insertion of HBOs in their brains, Greely identified a capacity that mice will not achieve, and then states that the important question is "whether you are creating something human-ish that you have to take seriously in terms of according it dignity and respect—and figuring out what that even means." The term "human-ish" obviously puts a born human and the HBO on the same spectrum of capacities.

Another ethicist quoted in the article asks, "if we give [NCs] human cerebral organoids, what does that do to their intelligence, their level of consciousness, even their species identity?" The journalist describes the problem of NCs as the possibility that "the animals might become too human" due to the human brain components.[61] Finally, a *Guardian* article says that HBOs could "gain consciousness, feel pleasure, pain and distress, and deserve rights of their own." Since only humans have rights at present, this too emphasizes that the HBOs could "become human." Greely states that, with enough complexity, "we potentially edge towards the ethical problems of human experimentation," which assumes that HBOs could obtain the protected status of humans in research.[62]

In sum, if we follow the public-policy bioethical debate, including the media reports citing experts from this debate, the primary or only reason to be concerned about HBOs and NCs is that if enough capacities are added to them, they could become morally human. Importantly from this perspective, there is nothing inherently wrong with creating an HBO or NC with human-level consciousness *if* we are prepared to treat it like we treat a human. But an HBO with human-level consciousness could not be kept in the prison of the petri dish, any more than a monkey with human-level consciousness could be confined to a cage. The tension in this research, from this anthropological perspective, is between making these entities enough like humans to be useful for medical research, but not so much like humans that we have to treat them as such, which would obviously preclude experimenting on them in the first place. As Greely notes: "if we make our models 'too good,' they may themselves deserve some of the kinds of ethical and legal respect that have limited brain research in human beings."[63]

We may wonder why bioethicists and scientists only use the philosophical anthropology to approach this issue, and why they focus on the problem of consciousness. I will largely leave this question about the source of expert knowledge to the side so as to focus on the public's stance on the issue.[64] Other reasons to support or oppose HBOs and NCs that are not incorporated into the bioethical debate will be enumerated below.

Conclusion

HBOs and NCs are developing rapidly. Each development will make the technology more controversial, despite the fact that the technologies are being developed with the most noble of intentions. I have so far described the technology and established that ethical policy must be set. This ethical policy is at present framed by the public-policy bioethical debate, which assumes that the main value at risk is that an entity could be harmed, and the only entity that could be harmed is one that has, at minimum, "consciousness." I have also established that all agree that the public's values should influence ethical policy; however, although ethical recommendations are already being issued, nobody has yet tried to determine what the public's values are on this issue.

In Chapter 2 I turn to examining the public by reporting what existing social-science and humanities scholarship says about the public's definition of a human and how that would impact its evaluation of entities like NCs and HBOs. In Chapter 3 I begin to report on my empirical research by describing the public's overall views of these technologies. In Chapter 4 I turn to the heart of the matter: whether and how the public's views of a human affect its evaluation of these technologies, and how these views differ from the public-policy bioethical debate. HBOs and NCs are instances of the broader category of biotechnology, and subgroups in society are more or less supportive of biotechnology in general. Therefore, in Chapter 5 I examine whether subgroups defined by their orientation toward nature, their attitudes toward scientific authority, and their religion have distinct views of HBOs and NCs. In the concluding Chapter 6, I examine what social-science and humanities scholars think is at stake in the creation of HBOs and NCs and propose ways to mitigate the risks they identify. I focus on the social slippery-slope problem: How do we take advantage of the positives of this research now without beginning the descent toward a dystopia that none of us want?

2

What We Know about the Public's Views of Humans

There are good reasons why the public-policy bioethical debate advances the value of avoiding harm and embraces the values of the academic philosophical or biological anthropologies.[1] But it is important not to assume that the current public-policy bioethical debate reflects how the public itself would actually see an issue—especially when experts are calling for the public to provide input to that debate.

This book will attempt to compare the US public's views of HBOs and NCs with those expressed in the public-policy bioethical debates. I limit this study to the United States because of methodological limitations and because I lack the required knowledge of other cultures. The methodology I use will allow me to make generalizations about all adult residents of the United States. Additional analyses of some subgroups will help to further develop my points.

In this chapter I examine the humanistic and social-science literature that discusses the public's view of humanity, but not specifically NCs and HBOs. The existing scholarship will help us interpret what I have found in my own study. I do expect that the public will share the positive reason for engaging in this research, which is the alleviation of human suffering. However, as we will see, existing research suggests that the public's concerns will be different from those of bioethicists.

Disembodied Brains. John H. Evans, Oxford University Press. © Oxford University Press 2024.
DOI: 10.1093/oso/9780197750704.003.0002

The Public's Definition of a Human

To understand the public's perspective on HBOs and NCs, we need to understand what the public thinks a human is (an anthropology) and how this differs from the academic definitions in the previous chapter.

A number of years ago I conducted an empirical study of what the US public thinks a human is. Let us start with some basics. The anthropologies of ordinary citizens do not perfectly map onto those of academics.[2] Moreover, whereas an ordinary individual may adhere to multiple anthropologies (while often tending toward one or another), academics see anthropologies as mutually exclusive. The public, unlike the academic community, is not rewarded for being logically consistent.[3]

When the various anthropologies were described in a survey, the one that garnered the most agreement was the theological, followed by the biological and then the philosophical.[4] The theological anthropology endorsed by the public was, at the level of detail necessary for this book, more or less the same as the academic version described in the previous chapter. One difference is that the soul was regarded as less like a means of communication than like one's "essence" or "true self." For example, it was common for respondents to say that the body is the container of the soul. As one respondent said, "a human is what God gives us to contain, like be a container on the earth, a home for the soul."[5]

In the philosophical anthropology used by the public, the particular capacities defining a human tended to be different from those of the bioethicists. The bioethicists tend to cite individualistic capacities like consciousness, rationality, and a sense of time. In contrast, the public tended to focus on more advanced *social* capacities of the kind that have also been identified by feminist ethicists, such as experiencing feelings, compassion, and the ability to make moral decisions and to communicate with others.[6]

The biological anthropology conceived by the public does not fit with the evolutionary continuum view of contemporary biology, but does fit tightly with the classic biological species concept and the evolutionary species concept, which emphasize human lineage and distinctiveness from animals. When the public talks about how biology makes us human, their explanations have three components. The first is that a human is an entity with a human body—an entity that looks like a human. The second is that a human is the offspring of two humans. The third is that a human always remains a human.[7] Though the public's definition could be called tautological, requiring a definition of a human to provide a definition of a human, it is probably better thought of as reflecting the older evolutionary species concept, in which humans are those who are in historical continuity with previous humans.

The main overall difference between the views current in the public-policy bioethical debate and those expressed by the public is that the public considers humans to be distinct from animals, which reflects the pervasive influence of the theological anthropology on US culture. Indeed, humans and animals are often not regarded as belonging on the same continuum. Even if the concept of the continuum is conceded, the public sees a sizable gap between animals and humans that makes humans special. This is evident in the survey from that earlier project which asked, "Which statement comes closest to your view about comparing humans to animals?" The possible answers were: "Are humans incomparably special, special, somewhat special, or not special at all compared to animals?" The first two options were selected by 73 percent, while only 10 percent selected "not special at all." That is, very few people view animals as having close to human-level moral status.[8]

To summarize, the definitions of a human used by the public emphasize human distinctiveness from animals, both physically and morally. This is reflected in the very prevalent theological anthropology that places humans on a different spectrum from the nonhuman animals, as well as a biological anthropology that emphasizes human biological distinctiveness. To a critic, this means that ordinary citizens believe in "speciesism," which "entails favoring members of a particular species (usually one's own) which is selected without naming any objective merits of that biological group."[9] The public's position could also be called "anthropocentric."

Moreover, the most commonly relied-on anthropology, the theological, is not materialist but instead assumes that we consist of a separable body and soul. Due to Christianity's influence on secular culture, the biological anthropology used by the public is similar to the theological anthropology; as we will later see, members of the public tend to implicitly rely on the theological anthropology, but in secular form.[10]

To get a fuller understanding of the public's anthropologies, let us see how someone who relies on the theological anthropology regards NCs. From this perspective, and in contrast to the concerns expressed in the public-policy bioethical debate, an HBO or an NC mouse cannot become a human either physically or morally. John D. Loike is a biologist who writes from the Orthodox Jewish perspective, which is similar to the Christian perspective. He concisely describes a view that I think most closely represents the majority view of the public. He writes that while secular bioethicists use "capacities based definitions" of personhood, "human status and personhood have different meanings. Biologically, the term 'human being' refers to an animal that genetically belongs to the species *Homo sapiens*. Many religious scholars include 'ensoulment'—defined as the instant a human being attains a soul—as a characteristic of human status."

In contrast to secular ethics, he continues, "any living organism that has human status also attains personhood status. But conferring personhood does not necessarily confer human status." That is, if a mouse obtains personhood, it does not become, morally or biologically (physically), a human. "Second, human status is given to any individual born from a human being and/or derived from human gametes regardless of its capacity-based functions or cognitive capabilities." This is the evolutionary species definition of the physical human preferred by the public, in which a human is the offspring of a human, combined with the morality in the theological anthropology. Therefore, "an animal that has been engineered with human brain cells or human neural organoids does not necessarily attain a human state. Being human requires the creation of an embryo from human gametes or being born from a human."[11] To take the most extreme case, a chimpanzee with a fully human brain, one that might stop you on the corner to ask about the weather and the stock market, is not human—although humans may grant it higher moral standing compared to other animals.

The Foundational Distinction between Humans and Animals

This section focuses on research that informs us about NCs. I hypothesize that the primary concern for the public is not the moral status of NCs per se, but rather that the existence of NCs violates the foundational cultural distinction between humans and animals. NCs are literally a cognitive threat to foundational beliefs about the world embodied in the public's definition of a human. The public varies in the extent to which it believes in this distinction; thus, those who believe more intensely in the distinction will be more opposed to creating NCs.

I am not the first to propose that this will be the source of public opposition. The most influential version of this argument in the literature is Robert and Baylis's claim that NCs will result in "moral confusion" for the public.[12] In their description of possible moral confusion, they roughly identify the public's anthropology, saying that "we explore the strong interest in avoiding any practice that would lead us to doubt the claim that humanness is a necessary (if not sufficient) condition for full moral standing."

More strongly, they write that violating this foundational cultural distinction will have negative consequences for society. They write that human-animal chimeras (neurological or otherwise) "threaten our social identity, our unambiguous status as human beings. . . . Hybrids and chimeras made from human beings represent a metaphysical threat to our self-image."[13] In a key passage they state:

> Indeed, asking—let alone answering—a question about the moral status of part-human inter-species hybrids and chimeras threatens the social fabric in untold ways; countless social institutions, structures, and practices depend upon the moral distinction drawn between human and nonhuman animals. Therefore, to protect the privileged place of human animals in the hierarchy of being, it is of value to embrace (folk) essentialism about species identities and thus effectively trump scientific quibbles over species and over the species status of novel beings. The notion that species identity can be a fluid construct is rejected, and instead a belief in fixed species boundaries that ought not to be transgressed is advocated."[14]

In the public-policy bioethical debate dominated by bioethicists and scientists, this perspective is very marginal, because it is not consistent with the anthropologies used by these professions. On the rare occasions when it is mentioned, it is not taken as a serious position that should influence policy but merely as something the public may believe.[15]

Foundational Distinctions in Studies of Culture

To proceed, we need to further understand foundational distinctions. Robert and Baylis did not invent this concept to use for NCs; rather, it represents an application of a long scholarly tradition in cultural anthropology. Anthropologists have long shown that societies embrace foundational cultural distinctions that are so deeply assumed that they are difficult for members of the society to even notice. These distinctions differ by society; in the contemporary West, some foundational distinctions are man-woman, mind-body, nature-culture, inside-outside the body, human-object, and human-animal.[16] Such distinctions have been most famously described by anthropologist Mary Douglas, who writes that the ban in Judaism on eating shellfish and bats is not the result of ancient wisdom about disease transmission, but rather that ancient Israelite foundational distinctions were based on these entities being categorically distinct.[17] A lobster lives in the water yet walks, and bats fly but are not birds. Cultural taboos are invoked when there is a blending of two foundational categories.[18] Thus, although Jews may eat both fish and beef, they are not supposed to eat lobsters.

You know you are looking at a violation of a foundational distinction in a culture when there is a taboo against bridging the distinction for which there is no other explanation, such as a mixing that produces something actually harmful. The mixes are often considered abominations, which are "anomalous or ambiguous with respect to some system of concepts."[19] For example, the Western taboo against necrophilia reflects a violation of the categories of living and dead; transgenderism was previously thought to violate the categories of man and woman; and incest violates categories of parent and child.

These foundational distinctions and their associated taboos cannot be easily discarded by the public. In one of my favorite examples of how deeply assumed these distinctions are, American anthropologist Matthew Engelke describes a visit to rural Africa during which his host offered him crickets to eat. Upon eating, Engelke vomited. In his interpretation, "I didn't throw up because I had a stomach virus. It wasn't a 'natural' or 'biological' reaction in this sense. I threw up because my body is cultural, or enculturated, itself. And in my culture we don't eat crickets."[20] For Engelke, crickets violated the American food/not food distinction. I hypothesize that NCs are generating a similar reaction.

Taboos from crossing distinctions, and the visceral reactions they engender, are stronger when they involve "us" humans. The philosopher of religion Jeffrey Stout sees a picture of a "cabbit," supposedly part cat and part rabbit, and calls it an "abomination" that disgusts him. But what would really disgust him would be an "anomaly with social significance," such as that which "straddles the line between us and them."[21] For Stout, "us" means humanity.

As you might have anticipated from my description of the public's anthropologies, one of the absolutely foundational distinctions in Western society is between humans and animals. Indeed, nearly all cultures make a distinction between humans and nonhuman animals, suggesting to some scientists that we have evolved to make this distinction.[22] This foundational distinction is indicated by disgust at bestiality. Bestiality might not hurt the human or the animal but is an object of disgust because it violates "one of the categorical distinctions most central to our society: the line between human and animal." Bestiality is not dangerous to those involved, but to society.[23]

To return to NCs, we know that chimeras violate foundational distinctions because they generate a visceral "yuck" response from the public. A recent study compared people's response to the different possible sources of replacement organs: mechanical organs, 3D bio-printed organs, human donor organs, and pig organs. It turns out that people really do not want the pig organs. "Pig organs invoke a 'yuck' factor," writes the author, and the organs express "concerns about pollution behavior, mixing up human and animal bodies, and blurring the boundaries between species." Using pig organs also "challenges known schemata of what it is to be a 'pig' and what it is to be a 'human.'" The author concludes that "respondents' views on xenotransplantation thus demonstrate concern both about policing species boundaries and—significantly—about protecting the individual's subjective identity."[24] So, as other scholars have noted, since chimeras mix humans and animals, we can expect opposition on those grounds alone. A possible research project to test whether chimeras really violate the distinction—a project unlikely to be approved by my university—would involve volunteers eating a chimeric pig, which they would presumably enjoy until told it was "part human." An important finding would be how many then vomit.

Belief in Foundational Distinctions Is Rational

It is important to not dismiss beliefs in foundational distinctions as emotional or irrational. Mary Douglas wrote the original study on this topic to show that such distinctions are found in both "primitive" and modern societies.[25] In the words of an interpreter of Douglas, "some central pairs of categories are widely shared: for example, man and woman, nature and culture, human and animal, organism and machine, life and death."[26] Moreover, the basic response of disgust is thought to be universal, selected by evolution because it was adaptive.[27] Therefore, "we need to take those categories and their resulting moral intuitions quite seriously, for we cannot do without our categories."[28]

Foundational distinctions are more powerful and consequential versions of cultural distinctions in general.[29] All distinctions are culturally determined, essentially chopping up the endless chaos of perceived experience to make it intelligible. Of course, societies—and individuals within societies—differ on the extent to which such distinctions are important. For example, Roma and Jewish cultures are particularly focused on maintaining distinctions.[30] I expect to find that members of the public vary greatly in the extent to which they believe in the human-animal foundational distinction.

The Jewish and Christian traditions may be particularly focused on making foundational distinctions. As Eviatar Zerubavel writes about Judaism (from which Christianity emerged), "No wonder this culture has thought up a purist God who actually spends the first three of only six days he has in which to create a world just making distinctions." Moreover, the Jewish and Christian traditions train adherents to make divides through ritualistic distinctions between the sacred and the profane, with particular rituals around sacred objects to make sure they are not contaminated.[31] On the other hand, Zerubavel also points out that at least the early Christians broke the distinctions of the time with their ethnic and social universalism, and they did, after all, believed in someone who was both man and God.[32]

Subgroups of a population can train themselves to not see particular distinctions that are otherwise dominant in a culture. I suspect that one reason why bioethicists and scientists do not write about the visceral reaction to violating the human-animal distinction is that they do not hold to this distinction. The scientists who create chimeras have been taught, with some effort, to ignore distinctions between life-forms that they most likely experienced as children. One summary of how biologists unlearn the human-animal distinction says that "The hybrid-generating power of the life sciences is rarely experienced by the scientists themselves as unnatural or disturbing because the techniques they use have long since been normalised within the field. This perspective is the result of years of training to seek knowledge in a particular manner."[33]

As one scientist in an interview study put it, "We all got used to the idea that life is a continuum. You can take something from yeast and put it in a mouse and something else from a chimpanzee and put it in a cow and more or less you can map things onto each other, and they sort of work." Another has said, "It's all very well to say people have been doing this [mixing] for forty years and every scientist knows about it. But outside of the biologists doing this kind of work nobody knows about it. It's never spoken of."[34] Bioethicists who focus in this area are likewise exposed to so much of this scientific work that it has been normalized for them as well.

Scholars who study foundational distinctions would conclude that people who claim they do not acknowledge any of these foundational distinctions and associated taboos see the distinctions that they do in fact observe as so natural that they are simply "fact." Creating distinctions discourages the self-reflection needed to recognize one's own prejudices. Indeed, "we regard any inquiry into the basis of the taboo as itself taboo."[35]

I suspect that the biologists reading this book decline to acknowledge a strong human-animal distinction, having been trained in a subculture that rejects it. They are therefore unlikely to experience a yuck response to seeing an NC. But I suspect all readers believe in many foundational distinctions and their associated taboos, such as the dead-alive distinction signaled by disgust with necrophilia. For the same reason, zombies probably also fascinate almost everyone. It is also hard to imagine that any reader lacks the inside/outside-the-body distinction. For example, piercings (which cross the skin barrier) are the object of both disgust and fascination, which are two sides of the same coin.

Finally, consider what your reaction to "plastinates" would be. Plastinates are whole human cadavers, often with the skin removed, sealed in plastic, and shown in museums. In one summary, viewers find them fascinating because "they transgress the familiar categories with which we usually make sense of the world: namely, interior or exterior, real or fake, dead or alive, bodies or persons, self or other. By refusing to occupy this familiar binary terrain, plastinates recreate tensions that our traditional concepts find difficult to encapsulate."[36] In sum, all societies recognize foundational distinctions, although they may be different in each society. Though people in the United States will vary, they generally believe in the human-animal distinction, and those who believe in it will, I hypothesize, tend to oppose NCs as threatening the distinction.

Will All NCs Violate the Foundational Distinction between Humans and Animals?

From the above discussion, we can be sure that if I created a horse with the head of a human and brought it to the local shopping mall to answer shoppers' questions, I would violate the foundational distinction and generate a very strong disgust response. If I created a mouse that stood up and said "Hi, I'm Mickey"—as in Greely's amusing example—I would get a similar reaction.[37] If that were what scientists are doing, then it seems quite evident that, for those who maintain a foundational distinction between humans and animals, seeing an NC would generate opposition to NC research.

We can expect a more muted and subtle reaction. As Peter Morriss points out, not all "blurring of boundaries is taboo." For example, "inviting animals into our homes, and families, as pets causes little disquiet."[38] Yet bestiality invokes the taboo, presumably either because sex is how new mammals are made or because sex is seen as part of the core of being human. What about an NC would represent a violation of the distinction? As we know, while using animal hearts to replace malfunctioning human hearts probably does generate disquiet, the reaction is not strong enough to lead people to actually oppose the practice.

The centrality of the components to the definition of the animal and the human that are mixed will matter. Violation of the distinction probably requires a mix of what is considered to be a core, defining feature of the human and a core, defining feature of the animal. By itself, inserting cells from a human into an animal will not ultimately invoke the foundational distinction. Instead, human appearance and behavior are probably key. As described above, the public's anthropology is based on not only the criterion of genealogy (born of a human) but also on the "looks and behaves like a human" criterion, which would be most likely to invoke a violation of the foundational distinction.

By contrast, in the philosophical anthropology used in the public-policy bioethics debate about NCs, what an entity looks like or does has little relevance to its moral status. In fact, advocates of the philosophical anthropology emphasize not letting appearance and actions sway decisions about moral status. A late-term fetus may look like a born human, as might a person in a comatose state, but what is important in the decision whether to keep either alive is the entity's capacities. Analogously, though a chimp or a dolphin may not look like a human, its mental capacities give it moral status.

A few bioethicists have speculated that the public would be most upset by the appearance of a chimera. While not giving moral weight to appearance, Greely has stated that "chimeras with some visible human characteristics could also be profoundly unsettling." I think he is right that people would be unsettled by chimeras with "human features, such as non-human primates with human faces or hands." He points out that this would "break the folk requirement of a strict separation between humans and other animals," and also possibly "create the kind of moral confusion" that is the concern of some scholars.[39] Similarly, according to a public consultation by the British government, people are particularly concerned with "cellular or genetic modifications which could result in animals with aspects of human-like appearance (skin type, limb or facial structure) or characteristics, such as speech."[40] Research into our reaction to the appearance of robots suggests the same.[41]

Along with appearance, behaviors are also key to triggering a violation. In my earlier study of anthropologies, the public did consider capacities to be part of their anthropology, but focused on social capacities like the ability to make moral decisions, communicate with others, have relationships, and feel love—that is, capacities that lead to behaviors.[42] Therefore, the strongest violation of the foundational distinction would come from endowing an animal with human appearance and behavior. To return to my dated cultural reference, Mr. Ed was fascinating for the public not because he had the requisite consciousness to think about his interests but because he could actually talk to Wilbur about them.

Whether the violation of the foundational distinction is weak or strong may depend on which animal is combined with the human. Animals that are evolutionarily farther from humans lack the basic appearance and behavior of humans and will likely produce less of a violation because the human qualities will not be expressed. One study of people's views of chimeras asserts that "some interspecies entities undermine the human essence whereas others do not. Here, non-human animals considered closer to the human, such as non-human primates, are argued to be unacceptable sources of biological material to be mixed with humans by a number of participants. . . . Non-human animals considered more different, such as rabbits and cows, however, do not constitute the same cultural risk to our sense of humanness."[43] This would make the dog less acceptable than the rat, because dogs are considered capable of all sorts of human emotions and reactions. Primates would presumably be considered even less acceptable.

Primates are particularly problematic because they already encroach on the foundational distinction, and adding human components would compound the threat. As Mary Midgley has pointed out, the hostile reaction to Darwin in the nineteenth century had to do with the views of apes at the time. To be equated with an ape was particularly insulting, because as "inferior imitators" of humans they were considered ridiculous. Physically they were also seen as abominations, since "by having a number of human characteristics, the ape was attacking our very conceptual framework . . . blurring the boundary between the human and the animal."[44] Apes were then considered evil or diabolical, as evidenced by "the association of apes with the Devil" and in artistic depictions of the Last Judgement "with ape-like creatures showelling the damned into hell." Morriss wryly writes that he suspects "that there would have been less opposition to evolutionism if Darwin had somehow managed to argue that we were descended from a noble animal like a horse."[45]

Armed with this sense of which NCs would lead to a violation of the foundational distinction, we can probably conclude that what scientists are *currently* capable of with NCs would not generate much of a response were the public to meet such an entity. For example, the monkey modified with one human gene would not behave differently or look any different. Even if the monkeys truly got smarter, this would be by monkey standards, and no one would say that they were capable of any humanlike interaction. The foundational distinction would be more threatened if the NC monkeys developed human behaviors, such as assembling machines in a factory.

In sum, *at present* the public is not going to *see* a NC animal and realize that it has a human component. But—and this is a key sociological insight—while the public responds to its own senses, its senses are mediated by our labeling. Therefore, even though someone could look at an NC mouse and think it looks like all other mice, the viewer could well experience a more limited sense of violation of the foundational distinction if there were a sign next to it that says "this mouse has a human brain in it" or more simply "human-mouse chimera."

Similarly, if violation of the foundational distinction does drive opposition to NCs, then naming the entity "part human-part pig" will remind people of the foundational distinction and generate disgust and opposition. An entity of this kind that has been given a made-up name like "Cebir" will less likely invoke the distinction in the viewer's mind.

I do not want to give the impression that the names given to these entities are all that matters for a perceived violation of the foundational distinction, or that the effect requires looking at an NC. The more general point is that talking about mixes of humans and animals will inevitably have that effect. In the passage I cite above from Robert and Baylis, they write that "asking—let alone answering—a question [about NCs] threatens the social fabric in untold ways." The more people read about an entity that is described as part human-part animal, even if they will never see the entity, the stronger the effect. Even reading this book would produce a tiny amount of the theorized effect.

To sum up this long section: I anticipate that, when I study the views of the public, I will find that the levels of consciousness of an entity are not very important to whether the public approves of HBOs and NCs, because the public is using a different anthropology than are the scientists and bioethicists. But I do expect to find evidence that opposition to NCs is driven by adherence to the foundational distinction between humans and animals. This opposition will become stronger if the qualities that matter for the public's anthropology, such as appearing like a human or exhibiting human behaviors, begin to be expressed in NCs.

Ephemeral Connections to Human Body Parts

I now turn to another likely feature of the public's anthropologies that is likely to produce distinct views of HBOs. Unlike NCs, HBOs are unlikely to ever be considered human by the public, because they will always lack the indicators of humanhood that the public recognizes. They will not be born of a human, they will not look like a human, and they will not act like a human. Greely makes the point: "the human tissues involved here, although they come from arguably the organ most tied to human identity, are small masses of disaggregated cells, suspended in fluid, contained in vials. An outsider looking at them would have no idea what they were. They do not have the more obvious humanity of a severed head, a skull, or a full skeleton."[46] And it goes without saying that, lacking a body, an HBO will never be able to engage in any of the behaviors that define us as human. So if violations of our sense of humanity and the foundational distinction between humans and animals are not at stake, what is?

Another foundational distinction is between humans and objects. We can treat objects however we wish. This distinction is indicated by the pervasive conversation about "objectification," which means treating a human like we treat an object. From the perspective of the public-policy bioethical debate, HBOs are objects. But if an HBO were somehow considered to be an ongoing part of a human, it could threaten the human-object distinction. How would that be possible?

The Materialist View of the Human

Let us delve into another aspect of both the philosophical and biological anthropologies used by bioethicists and scientists in the public-policy bioethical debate about HBOs and NCs. These two anthropologies assume that the human is *only* matter. "Material" or "matter" essentially means "made from atoms," and the belief that nothing exists besides matter is called "materialism." The body is made of cells, and cells are ultimately made of atoms.

As a result, for the materialist, the mind and consciousness are ultimately physical, reducible to chemicals in the brain. In one account of this position, "everything that exists, including all mental states and properties, is entirely constituted by matter arranged in particular ways."[47] A classic materialist statement comes from the codiscoverer of the structure of DNA, Francis Crick, who said that "you, your joys and sorrows, your memories and your ambitions, your sense of personal identity and free will, are in fact no more than the behavior of a vast assembly of nerve cells and their associated molecules."[48] There is then no "spark," there is no "soul," and all of mind and consciousness is part of the human body. The mind, its thoughts, and the body are all one piece.

Materialism, when applied to the human, "is the idea that human beings are material objects—brains, perhaps, or some part of the brain—without immaterial selves or souls."[49]

Nonmaterial Part of the Human

In opposition to materialism (and monism more generally) is *dualism*, which holds that there is a distinction between body and mind. I am interested in the version of dualism in which there is a body made ultimately of material (atoms), and another feature of the human that cannot be reduced to material and the body. As one conservative Protestant philosopher puts it, "according to the most popular form of dualism—one embraced by Plato, Augustine, Descartes and a thousand others—a human person is an immaterial substance: a thing, an object, a substance . . . and a thing that isn't material, although, of course, it is intimately connected with a material body." He wants to defend the idea that "a human person is not a material object."[50]

Part of the public's theological anthropology is that an existing born human has a soul, given by God, that is not material. This soul resides in your body until, at death, it goes elsewhere. In one interviewee's summary, "A human is a man or a woman created in the likeness of God that has a spirit, body, and mind, and a heart." According to another interviewee, "We are a spiritual being that's having an earthly experience, or human experience. So we have the human body . . . [and] we have the spiritual side."[51]

It is not only Christians who believe this. A lot of religious ideas become secularized over time, losing their explicitly religious content but retaining the basic form. My earlier study of the public's view of the human shows that many of the people who were not relying on the theological anthropology nonetheless had a notion that people have a nonmaterial soul. A minority held to a materialist notion of the human, such as one respondent who, in response to the question of whether there was a soul, said, "I don't think so. I believe an integrated system of how the brain comes together to form information in the brain. . . . It's not just neurons that have recorded information; it's how the information is drawn from the neurons . . . that kind of electrical activity." Some thought of the soul as the self—a kind of psychological construct.[52]

Others thought that the soul, more analogously with the Christian soul, was "energy," with one respondent saying, "The soul is an energy that exists with us and allows us to be connected to each other and allows us to be connected to nature and things that we don't quite understand." It was also common to think of the soul as a person's "essence": "It's this kind of soul or essence that is distinct from just our body and our brain. . . . Your soul can live on after your body is gone."[53]

This belief that humans have a soul can lead to believing that a person can retain a connection to his or her disembodied human parts. The soul and the body are normally integrated, but if the soul can travel and a part of the body goes elsewhere, the soul could connect the body and the part. This is important because HBOs are disembodied human tissue that were grown from the cells of a particular human. I will call such linkages "ephemeral connections."

In contrast, for bioethicists and scientists HBOs have no moral status, for two reasons. First, per our earlier discussion, HBOs lack the capacities for personhood. Second, HBOs have no physical and material connection to an actual person. They have literally been "objectified," made into an object, with a status akin to that of a chair. This view is exemplified by doctors removing diseased parts of the human body and unceremoniously throwing these parts away. For those who have a materialist view of the human, it is incomprehensible that people would retain a connection to their disembodied parts—and indeed this possibility is not mentioned in public-policy bioethical debates about HBOs.

Ownership as an Ephemeral Connection

There *is*, however, one very narrow connection between a human and his or her disembodied parts that is recognized by bioethicists: ownership. You own, or at least control, the parts of your own body. This is a more general issue in that it is not specific to HBOs or NCs. Viewing the connection between humans and their parts as ownership or control results in a particular version of ethics in which an act is acceptable if consented to.

For example, Henrietta Lacks was a poor African American woman whose cells were taken without her consent, replicated, and sold as the basis of the HeLa cell line used in biomedical research.[54] From a materialist perspective, the primary issue here is that she owned those body parts, did not give consent for them to be used, and did not receive the profits. If she had agreed to give away her body parts, then any connection to those parts would have ended. Consent to release ownership over body parts, typically for donation, is thus a critical issue for bioethicists.

The Public's Ephemeral Connections to Body Parts

While the public would likely agree that they own and should control their body parts, they are unlikely to agree that giving a part away severs all connection to it. Consider organ donation, which is premised on the idea that we own or control our body and that, once the organ is given, there is no longer a connection. Indeed, the language of the "gift," used in organ and blood donation, suggests that, once given, the organ is no longer related to the giver. But, as has been noted by many an anthropologist as well as scholars who study organ donation, a gift creates a relationship between the giver and the recipient, setting up an expectation of reciprocity.[55] In the words of one analyst, "gift theory explicitly poses some degree of continuity between the giver and the gift. This suggests that tissue donated to another does not conform to a strict property model but is inflected through more diffuse and less defined transactions. . . . Tissues detached from the body do not easily lose their designation as 'self.'"[56] The relation between owner and gift recipient or buyer is the most recognizable ephemeral connection with human body parts. To understand how people may come to view HBOs, we need to focus on even more ephemeral connections to disembodied human parts.

I suspect the most widely read text in the ethics of science and medicine is *The Immortal Life of Henrietta Lacks*, which I mentioned above. This book has many messages about racism and power, but for my immediate purposes the book is also a report of how at least one family and those around them viewed disembodied human parts. The book not only reports these views, but as a major best-seller it continues to teach its many readers a particular view of the body and its disconnected parts.

The family and those sympathetic with their story do not exemplify a materialist attitude toward disembodied parts. From the materialist perspective, those cells may have started in Henrietta Lacks, but, once removed, they have lost any connection to her. Yes, the book makes clear that she or her family should have been asked permission to surrender her ownership rights. But the book's title suggests that the deceased actual human, Henrietta Lacks, has become immortal because her cells live on. She retains a connection to those cells now growing all over the world in labs.

The book's author reports that when Lacks's daughter found out that "scientists had been using HeLa cells to study viruses like AIDS and Ebola, [she] imagined her mother eternally suffering the symptoms of each disease." In another scene in the book, family members are shown a freezer full of HeLa cells. One of Lacks's children says, "I can't believe all that's my mother." When handed a vial of the frozen cells, Lacks' daughter says, "she's cold." At another point the daughter, looking through the microscope, says "This is my mother. Nobody seem to get that."[57] While the daughter does not use the term "soul," clearly she believes her mother's essence is in these cells.

In a scene that suggests why Christians may be particularly supportive of the idea that we are not only matter, another family member says, "He [Jesus] died for us that we might have the right to eternal life," and continued, "You can have eternal life. Just look at Henrietta." The reporter asks: "You believe Henrietta is in those cells?" and he replies, "Those cells are Henrietta." He goes on to assert the common Christian dualism that there is a physical body and a spiritual body. The reporter asks, "You're saying HeLa is her spiritual body," to which the family member smiles and nods.[58] Note that these disembodied cells are not even the original cells, but were grown from those cells, yet still "are" Henrietta. If the cells had been made into 1,000 HBOs, the family would evidently have the same reaction.

Beyond the case of Lacks, there is a social-science literature on these ephemeral connections that will help us understand the public's view of HBOs. The literature depicts both weak and strong versions of ephemeral connections between the donor and his or her disembodied tissue. I start with the weak. In a study of the public's view of blood—which I would argue represents less of a human connection than does the brain or heart—sociologists asked blood donors if they thought the blood was still "theirs" after donation. About half the sample had the ownership view, which was that blood is "alienable" and "once donated, no longer referred to the donor in any meaningful way."[59] Other respondents expressed the weak version of an ongoing ephemeral connection, which the authors call "continuity of possession," with one respondent saying the blood "would be always mine. . . . It will always stay my blood," resembling a perpetual loan. Still other respondents saw the connection as like that of parent and child, saying "It's like children, they may grow up and be their own person but you're still my child."[60] This is less like a property relation and more like a family tie. The public may see a connection between an HBO and the stem-cell donor akin to parent and child, with that connection to an actual human influencing their opinion about the ethics of creating HBOs.

The strong version of the ephemeral connection suggests that the soul of the person still inhabits the parts. For a small group of the respondents in the blood-donor study, blood was a "bearer of personality or moral disposition" that would be transmitted to the recipient. The human tissue also conveys "the moral disposition of the donor to the recipient," which led some people to say that if they received a blood donation, they would like it to come from a good person.[61]

Similarly, in a focus-group study of tissue specimen donation for research, the authors heard a number of concerns recognizable to bioethicists, such as loss of privacy. But they also encountered another category of resistance, which held that "biospecimens are an extension of the 'self' and require special care and respect." For those who talk this way, "the physical nature of biospecimens retains an individual's essence." As one of their respondents said, "I feel more uncomfortable with giving a biological specimen because it feels like an attachment or extension of myself."[62]

The strongest ephemeral connection is the perceived transferral of the self of a dead person via transplanted organs, a phenomenon that has been recognized since the earliest days of transplantation. In an empirical study of organ donation, the authors found that the "recipients seem to share the belief that some of the psychic and so-cial, as well as physical, qualities of the donor are transferred with his or her organ into the person in whom it is implanted. . . . The sense that part of the donor's self or personhood has been transmitted along with the organ is likely to be most pronounced with cardiac transplants. Although on the surface recipients regard the heart they have received as just an organ . . . on deeper levels many re-spond to it as if it were a repository and emanation of the donor's quintessence."[63] Organ recipients thus worry that they may be influenced by the identity associated with the organ.[64]

Another summary of these studies states that "[c]hanges in gender or behavior, a resurgence of youth, or discovering new tastes and preferences are often attributed by transplant recipients to the act of taking an organ from one person and transplanting it into another."[65] Another study found a divide between seeing the organ as "nothing more than a 'machine part'" versus "an animate or spiritual extension of the donor."[66] Other anthropological studies find that the recipients of organs often feel as if they have the memories of the (often dead) donor.[67]

As with all social-science research into topics such as this, we need to be reminded that these effects are tendencies rather than certainties. Few people will categorically believe that an HBO has retained a connection to the donor; instead, some will have an inkling or a suspicion that nudges them in this direction. The authors of the blood-donor study found this ambiguity in their subjects, and the words of one of the respondents are worth quoting in full: "My immediate response was the minute [donated blood] goes out of my body it's no longer mine because . . . I've consented to make it somebody else's. . . . But I guess in another sense it's always mine because it was kind of manufactured by my body and it's a tissue of my body that's going to be transplanted into somebody else's so it both is and isn't my body's simultaneously."[68]

The body part that houses the soul should have the most powerful ephemeral connections to the original human. Organ-donor studies have shown that the connection is strongest when the donated organ is the one that (apart from the brain itself) is most representative of the self—that is, the heart. As one of the medical professionals involved with organ transplantation wrote, "Donor families think that when they donate something, certainly the heart—the loved one lives on in some way."[69]

In my interviews for this book I spoke of the donated cells used to create the HBO as coming from someone's skin. In popular accounts, the skin does not contain the soul. However, in the laboratory the skin cells grow into a "brain." The soul leads to communication with God and to good and bad behavior, and thus is your "true self," and this all would reside in the brain. So the connection of an external brain to the existing human should be strong.

To tie this all together: I would expect that a subset of Americans thinks that HBOs have one of these ephemeral connections to an actual human, and that this will make these people more wary of creating an HBO because it would somehow negatively impact the donor, or would even "be" the donor, now trapped in a petri dish.

I earlier wrote that, for an NC to violate the human-animal distinction, more than viewing the NC would be required. Similarly, merely observing an HBO will reveal nothing; nobody can look at an HBO in a dish and know that it was grown from the cells of another human. Rather, the impact on our views of the human will occur when the HBO is labeled or discussed as "grown from another human's cells," or simply "human."

Conclusion

When NCs and HBOs emerged in the news, the reporters who understandably contacted the ethics experts were told that what was most important was whether an NC or HBO could attain human-level consciousness. But these experts have used particular definitions of the human that the public does not fully share. If we examine the public's definition of a human, we see that nonscientists make a very sharp distinction between humans and animals, which has become a foundational cultural distinction. For the public, opposition to NCs is less likely to be due to concerns about the NC's consciousness than to the perception that the NC is violating this foundational distinction, which in turn threatens our own sense of ourselves.

The human-object distinction is also threatened because the HBO as object is considered part of the human. In addition, the public's definition of a human, unlike the definition used in the public-policy bioethical debate, is not fully materialist. Humans are bodies, but they are also souls—souls that can depart from the body. Other studies have shown that, because of this more spiritual conception of a human, the public will tend to sense ephemeral connections to human body parts. The public can be expected to assume that HBOs retain a connection to the human who donated the cells, and this connection will likely make them less supportive of HBO research, because it treats an HBO like an object, and probably also because of fear of somehow harming the cell donor. With this broad background in place, I will turn in the next chapter to investigating what the public really thinks about NCs and HBOs.

3

The Public's View of Human-Brain Organoids and Neuro-Chimeric Animals

The previous chapter examined the humanistic and social-science literature to identify what to look for when examining the public's views. We now turn to that examination. These abstract humanistic questions are difficult to evaluate within the standards of sociology, and, because of the limitations of social science, I cannot empirically evaluate all the questions raised in the previous sections. It is much easier to ask someone "What is your occupation" than "What is your definition of a human?" Therefore, my analysis does not give precise answers, but is suggestive. I will present many pieces of a puzzle which in the end will not all be tightly linked, and there will be a varying amount of space between the pieces, but they will still allow us to determine that the puzzle is of a picture of a sunset and not a bear.

Ideally, this research would be based on existing social-science studies that are directly focused on HBOs and NCs. However, as you would expect, there is little social-science research on HBOs or NCs in the United States because the technology is too new, and what does exist is not relevant to the questions being asked in this book.[1] Our study will partly depend on adjacent empirical literatures, on subjects such as the use of animals in research and the public's views of nonneural chimeras.[2]

Like many studies of emergent technologies, the existing social-science studies generally focus on measuring the percentage of

Disembodied Brains. John H. Evans, Oxford University Press. © Oxford University Press 2024.
DOI: 10.1093/oso/9780197750704.003.0003

the public that approves of a technology, and sometimes on which demographic groups are more supportive than others. Their overall conclusion is that the public has mixed views, and the percentage of the public supporting a technology will be highly dependent on how the survey question is worded. Thus, we will focus on questions about *why* someone would be supportive or opposed to HBOs or NCs.

I will primarily depend upon a nationally representative public-opinion poll. (For details about the survey, see the Methodological Appendix.) The price paid for representativeness is the inevitably thin results that come from asking respondents if they agree with 10-word statements. However, we will leaven the necessarily deductive and thin data that such surveys produce with some richer, qualitative, in-depth interview data.[3]

The survey employs two types of questions. The first are the standard attitude and demographic questions seen in sociological surveys, such as, "To what extent do you agree with X?" The second are experimental vignette questions. This chapter will describe the vignettes carefully, since they will describe the public's overall view of HBOs and NCs and provide the basis for many subsequent analyses. Subsequent chapters will dig much deeper and ask *why* the public has these views.

Experimental Design Vignettes

In a standard survey, the questions identify relevant characteristics of the respondents, such as their religion and their attitudes toward HBOs. A typical analysis would then infer that, if, say, conservative Protestants are less supportive of HBOs than are the other respondents, then it is likely their conservative Protestantism that has led to their view of HBOs.

But the true cause of a given person's views could actually be a different characteristic which happens to be correlated with both religion and views of HBOs. That is, religion may actually be a proxy for some other characteristic that members of that group tend to share. In this hypothetical case, it could be that Republican Party identification is the characteristic that actually leads to views of HBOs, and that conservative Protestants just tend to be Republicans. This problem of interpretation is typically addressed by *controlling for variables* in statistical models.

For example, surveys show that conservative Protestants are less likely to believe in climate change than other Americans. However, after controlling for party identification, the party-identification characteristic reveals itself as the more precise cause—and the conservative Protestant characteristic is not associated with views of climate change. The relationship between conservative Protestantism and beliefs about climate change is not due to conservative Protestantism per se, but instead reflects the fact that conservative Protestants are disproportionately Republicans.[4]

The problem is that you never know if you are controlling for all of the right characteristics—and indeed the critical characteristic may not even be measurable. Surveys in a still undefined area such as attitudes toward HBOs and NCs may potentially present an even bigger problem, because we really do not yet know what to control for.

One solution is to use an experimental vignette design. In the social sciences, a vignette is a short description of a situation, in which small components of the description can be experimentally manipulated to create different versions, and respondents can be randomly assigned to read particular versions. Since the assignment is random, there is no need to control for the respondents' other characteristics.[5] For example, if we were interested in whether people think women and men should receive equal pay, a vignette could be created describing a worker's job and then asking how much that worker should be paid. If the vignette contains a number of details that the researcher is not actually interested in (hours worked, occupation, and so on), the respondent will be unaware of what the survey is actually looking for. But suppose that half of the respondents randomly see a version in which the person being discussed is named "Linda," and half see a version about "Mike," and after the vignette they are asked, "How much per year should the person be paid for this job?" If those who were randomly assigned to see "Mike" suggest higher pay than those who saw "Linda," this is evidence for discriminatory attitudes toward pay.

To return to my case: Using a standard survey approach, respondents could be asked what value, from 1 to 10, they would put on an HBO that possesses the consciousness of a human fetus, and then asked whether they support the creation of HBOs. If the two answers were correlated, we could say that those who are more concerned about entities with human-fetus-level consciousness are more opposed to HBOs. However, there are many variables that could be correlated with these two that could interfere with this interpretation. Men may generally be more opposed to HBOs, and also more likely to devalue something with the consciousness of a human fetus. If so, the seeming relationship between consciousness and approval may actually simply reflect the fact that men are more opposed to HBOs and have a particular view of consciousness. While gender can be controlled for statistically, other characteristics could be harder to identify and account for.

An experimental vignette can avoid this problem because respondents are randomly assigned to see the vignette in which an HBO is described as having the consciousness of a human fetus, and thus level of consciousness cannot be correlated with the characteristics of the respondent. This is the exact same logic as a randomized clinical trial for a vaccine, in which people are randomly (and secretly) assigned to receive either the drug or a placebo. If those who received the drug get less disease than those who received the placebo, we know that the drug's effect was *not* due to the drug recipients being more female than the placebo group, or more educated, or prone to some different behavior, because the drug was randomly assigned. The same is true for the vignettes in this survey. Therefore, if we see that people are less approving of NC research involving monkeys compared to mice, we can be certain that this is not actually the result of some demographic group disproportionately liking monkeys and simultaneously disliking NC research.

Another advantage of a vignette is that it is typically, by survey standards, a long description. This does not matter for simple questions like "Do you have children," where all the words are crystal clear to respondents. But when few participants will know what an HBO or an NC is, the vignette allows the researcher to provide basic background information.

There is an even more important quality of experimental vignettes compared to standard analyses. In a well-designed vignette, the respondent cannot determine what the person who wrote the survey is looking for. To continue my gender-and-wage example, the respondents do not know that the analyst is comparing men and women, because they only see one of these two possibilities, so there is no response bias in favor of perceived social desirability.[6] If those who saw "Mike" on average propose a higher salary than those who saw "Linda," then those respondents overall think that men should be paid more than women for the same job. By contrast, if asked outright about equal pay for males and females, or asked two separate questions about what Mike and Linda should get paid, most respondents would likely choose the socially acceptable response of equal pay.

Returning to my case: If a standard survey asked three separate questions about the acceptability of creating HBOs with the consciousness of an insect, a pig, and a human fetus, respectively, respondents would likely be most opposed to creating HBOs with the consciousness of a fetus, because they would be comparing the three questions. Indeed, evidence from the in-depth interviews excerpted below suggests as much. The respondents would realize that the investigator is implicitly ranking those questions, and if they care about consciousness (as the questions themselves imply they should), they will be most opposed to the HBO with consciousness at the human-fetus level. In the vignette, by contrast, they are presented with only one of those possibilities, so they have nothing to compare it to and cannot know the motive of the survey researcher.

Given the benefits of experimental vignettes, why would anyone engage in standard survey analysis at all? The answer is that vignette analysis can only be used to examine a few research questions without running to the length of a Dostoyevsky novel, with all of the attendant attention and comprehension problems. (This will become clear in the following pages.) For more on the technical issues presented by vignettes, I refer the reader to the Methodological Appendix.

The HBO Vignette

My conclusions about people's views of HBOs will depend on how HBOs are described in the survey. Since vignettes must be very short, lacking space to describe all possible futures for these technologies, there will be room for only a few exemplary possibilities. The description cannot be limited to imminent developments, because the impossible has a tendency to become the possible, and usually more quickly than we think.

In selecting a description of an HBO, I chose scenarios similar to those the public will hear when news of HBOs eventually spreads. As soon as the public hears about these technologies, the stated concerns of scientists and bioethicists can be expected to be front and center. The technologies will be described as beneficial for medical research, and potentially life-saving. Since that is how the public will encounter these technologies, that is the framing I will use. If advancing human health was not mentioned, the data would obviously indicate far more opposition.

We can begin with a discussion of the vignette's design. (This will become particularly important in the next chapter, so please bear with me.) The concept that randomly changes in a vignette (e.g., the person's gender in the employee salary example) is called the *factor*. The value of the factor (e.g., female) is called the *level*. When reporting on a vignette in an academic publication, the factors are demarcated by brackets ([ ]), and the levels within the factor are separated by a slash (/).

Below I break down a sample vignette, underlining the phrases representing key concepts in the level.[7] In my survey, the respondents saw only one level of each factor, without underlining, and the text was strung together to produce a short paragraph. The vignette was introduced as follows: "We now would like your view on another new scientific development. Please read this carefully because we will be asking specific questions about it." It then began:

A human brain organoid is a [1/2 of an inch / 5 inch] piece of a human brain.

I selected this factor because in science and bioethics the size of an organoid is irrelevant; all that matters is its capacities. But in the anthropology that the public uses, a larger organoid may invoke similarity with humans. Again, a randomly assigned 50 percent of respondents saw ½ of an inch and the other 50 percent saw 5 inch. The question is whether those who saw 5 inch react to HBOs differently from those who saw ½ of an inch.[8]

The vignette continued, with 50 percent reading that only one HBO was created and 50 percent reading 10,000. Half of respondents saw that HBOs were "grown" and half saw that they were "manufactured." Thus, combining these two factors, 25 percent of respondents saw:

[One was grown] by re-programming stem cells from the skin of the scientist doing this research. It is[9]

By ensuring that some respondents were reading about mass man-ufacture of HBOs and others were reading about a single instance, the levels in this factor are designed to invoke differential senses of human instrumental control of nature. (This will be explained in Chapter 5.) The vignette then continued with a bit of text with no variant language, which all respondents saw:

> kept in a dish. An organoid can be connected to muscle and eye cells, and can receive electrical input. One question is whether, if organoid research continues, an organoid could achieve

The "kept in a dish" introduces one of the key contextual points—the fact that these HBOs are not walking around. This is followed first by the key point that HBOs can interact with other cells, and then by the lead-in to achieving consciousness, which is the next factor.

The next phrase (following "could achieve") introduces the final complicated factor, which is the capacities of the HBO. Half the respondents saw "self-awareness—knowing that it exists," and half saw "awareness of its surroundings—know that things are around it." This distinguishes between self-awareness—long thought of as the highest and most humanlike awareness—and the situational awareness that a mouse may have.[10] This factor is an attempt to measure a subtle distinction that is nevertheless very important to scholars concerned with consciousness.

Experts could object that awareness is not the same as conscious-ness. However, given that there is "no aspect of mind more familiar or more puzzling than consciousness" and given "the lack of any agreed-upon theory of consciousness," "awareness" was as close as I could get in the space available.[11] Six- to 10-word summaries in a vignette obviously cannot stand in for entire theories of consciousness.

The respondent was also assigned to see a description of the awareness. The levels were: "will not be aware," "as an insect," "as a pig," and "as a human fetus."[12] These test the assertion in the bioethical literature that evaluation of the morality of an HBO is dependent upon its own level of consciousness, with the perceived consciousness being ranked in the order of the four options above. Again, as a researcher I'm hoping to learn whether people who see "as an insect" are more supportive of HBOs than those who see "as a human fetus." The vignette finishes with a sentence that all respondents saw:

> Scientists want to experiment on these organoids to develop treatments for human brain disorders such as Parkinson's or Alzheimer's disease.

So, for example, one of the 64 (2×2×2×2×4) possible vignettes that a respondent could have been randomly assigned to see (without underlining) was:

> A human brain organoid is a 5-inch piece of a human brain. One was grown by reprogramming stem cells from the skin of the scientist doing this research. It is kept in a dish. An organoid can be connected to muscle and eye cells, and can receive electrical input. One question is whether, if organoid research continues, an organoid could achieve self-awareness—knowing that it exists. The conclusion so far is that an organoid could have as much self-awareness as an insect does. Scientists want to experiment on these organoids to develop treatments for human brain disorders such as Parkinson's or Alzheimer's disease.

Note that I have framed this vignette using the terms of the public-policy bioethical debate, so respondents would think that the debate is between advancing human health and dealing with the potential consciousness of the HBO. If I had framed this as a debate between human health and the source of the original human stem cells (hand, heart, brain, cadaver, adult, baby), I would obviously obtain different levels of approval, and different groups would be supportive or opposed. In the next chapter we will see if academic distinctions such as consciousness register with the public.

General Reaction to HBOs

It is also important to get a sense of the public's reaction to HBOs beyond clicking on Agree or Disagree statements in the survey. Examining the in-depth interviews, we notice a disgust response. There is also a sense of disbelief that such technology is underway, with respondents invoking science-fiction references. For example, our first interviewee, Jane,[13] reacted to the possibility of an HBO becoming aware by saying:

> I don't know. It just starts to sound a little too much like playing God, but what if it became a real thing? I don't know. That would be a little creepy, because then, what do you do with it if it's a real thing and got out of control, something like a bad science-fiction movie?

When I asked Michele about the consciousness of an HBO, she said, "This is where it gets super-creepy. I would say that . . . it's still . . . it's weird." Similarly, Vanessa said, "Whoa . . . it's like that book . . . *My Sister's Keeper*, where they had the second daughter, so that she could give all her blood transplants and marrow transplants to her sister with cancer, and that was like her whole reason for being alive is to be an organ donor for her sister." Isabel sees HBOs evolving into a "more, like, sci-fi situation." Steve said he wanted a set of rules established for HBO research, "so we don't turn into Frankenstein."

The in-depth interviews reveal the extent to which HBOs flummoxed the respondents. The public typically reasons by analogy, and clearly the respondents could not think of anything HBOs were analogous to. (As we will later see, respondents had an easier time thinking through NCs.) Thus, the interviews were full of pauses, repeated words, restarts, and half-finished sentences in which the respondent would decide to start over and say something differently. Isabel's discussion of one of the vignettes was full of hesitations and reversals (unlike her response to in-depth interviews on other topics): "I mean, I guess, like, good. Enough human qualities, like, at what level? Like, if these . . . I mean, disagree. I think. So stressful. . . . Yeah, I don't know, like the, like the fulcrum at which I would be, like, actually. Yes, these are . . . Man, I'm . . . sorry. My brain is just running through a lot of scenarios!" Another interviewee, Brian, described his own difficulty somewhat more coherently: "Yeah, that's what I'm kind of trying to puzzle out. I'm sorry I got real quiet there. You could hear the marble rolling from one side of my head to the other, through its little maze there."

This befuddlement suggests that the public's current views may not prove to be very durable. It is possible that, 10 years from now, the public will have been taught ways of thinking and talking about HBOs by reporters and commentators, just as they have by now learned the "pro-choice" and "pro-life" language used in the abortion debate. But for now, the public lacks established ways of thinking about these issues. The first take on these issues by members of the general public, as recorded in my interviews, will likely turn out to be representative of the broader public's views when knowledge of these technologies first becomes widespread.

Survey Responses to the HBO Vignette

To return to the survey: Immediately following the vignette, the respondent was asked to evaluate the statement "Research into human brain organoids should continue," using a five-point scale of responses ranging from "Strongly agree" to "Strongly disagree," which was coded for the statistical analysis so that "Strongly agree" received the highest number. I consider this to be a general statement about approval of the research.

The frequencies of selecting particular responses are an average across all the factors in the vignette, again made possible by the levels being randomly assigned to the respondent. It turns out that there is very little opposition to the research as I have described it. Only 4.3 percent of those polled strongly disagreed that the research should continue, and 4.8 percent somewhat disagreed. 21.4 percent selected "Neither agree nor disagree," 41.5 percent selected "Somewhat agree," and 28.0 percent "Strongly agree." Thus, there is evidently very strong support for the research, which is in and of itself an important finding. For most people, the negative anxiety regarding the creepy "brain in the dish" does not outweigh the positive appeal of medical research.

We can use the in-depth interviews to get a sense of the trade-offs behind clicking "Disagree" or "Agree" on the survey.[14] A strong majority of the respondents in the survey are approving or somewhere in the middle, and they see two sides to the issue, eventually coming down on one or the other. For example, Mark, comparing his response to a previous answer, says of the HBO vignette, "I'm gonna go with just "Agree" this time instead of "Strongly agree." I'm definitely not as confident." I then asked, "Where's that shaky confidence rooted?" He replied, in the typical, somewhat flummoxed manner of other interviewees:

The fact that the, that the sample could possibly become self-aware. And I think . . . having like . . . yeah, I mean it is human brain tissue. I don't know how fully developed it is, if you can . . . if it's . . . like, that's definitely a gray area for me that there's a self-aware brain organoid in a dish.

I responded, "So, given all those caveats, you still agree that research should continue. What's good about this, then?" He answered:

If this research could be pursued in an ethical way, I mean, . . . There could be huge benefits to a lot of people who suffer from Parkinson's and Alzheimer's. I see the potential to help a lot of people. Yeah, I guess it would depend on if there's a way to measure if these brain organoids are able to have self-awareness and try to avoid having them reach that state, if you could avoid that, then I would agree that it should continue, but if you're getting to a point where they're becoming self-aware, then I, then I think, maybe pump the brakes a little bit and kind of real fast.

Dan was typical in seeing the downside referenced in the vignette's consciousness discussion, but believed so strongly in the medical goals that he strongly agreed with continuing the research:

My default response, it would be like to be afraid of the potential of these things. But in the scope of, like, fear, this is an existential risk of these things, actually. There are definitely more moral questions that I'm not thinking about, like, if these things can perceive its surroundings. Can it feel? Would it be in pain? But I don't know—we just have to find out. So, you have to venture, and there are risks to venturing into the unknown. And autism [and] senility like . . . Alzheimer's. Yeah, that sucks—that's one of my biggest fears.

It is then no surprise that the respondents most approving of the research were those who were motivated by the relief of suffering, and those for whom disembodied human tissue and any potential level of consciousness of the HBO was unproblematic. For example, the issue of consciousness in the vignette did not really register for Isabel, who concluded: "I mean, mostly it seems weird and cool, which is great. Obviously, like, Alzheimer's and Parkinson's research is, like, very important. My family has a ton of Alzheimer's history, so I think . . . let's fix that before I become old and decrepit.

Yeah." Similarly, Bill said, "When I look at those two examples, specifically with Alzheimer's and Parkinson's, it, it's strongly in the neurological department. . . . I have extended family so it's always on my mind that, you know—gee, it'd be great if we could do something here."

I cite these examples from the interviews to give us a better context for understanding the numerical survey data. People are torn over issues such as this, and even most of those who come down on the approving side at least understand why someone else might be more hesitant. In the following pages I will be describing what sort of person ends up where on the disapproval-approval scale. Meanwhile, it is important to remember that the thought process leading to one position or the other can be involved and complex.

The Neuro-Chimera Vignette

Let's now turn from HBOs to NCs. In investigating respondents' attitudes toward NCs, I again employed the vignette technique. As in the previous examples, the critical concept in each level of each factor will be underlined.

The NC vignette was introduced like the HBO example: "We have a final scientific development to describe to you. Please read this carefully because we will be asking specific questions about it." It then begins:

> Animals are experimented on in medical research to develop cures for human disease.

I included this sentence to lay out the moral issue here, which is that such experiments are intended not simply to satisfy scientists' curiosity but to relieve the suffering of disease. The vignette then continues:

Scientists plan on creating [no more than a dozen / a new sub-species of][15]

As in the HBO vignette, the intention here was to distinguish between modifying a few animals (which we do all the time) and creating or designing a new life-form in nature, obviously a more interventionist act implying a particular view of the human relationship to nature, which I will examine in Chapter 5.[16] The sentence continues:

[mice / monkeys] that have been

Mice were selected as an example because they are the current animal of choice in laboratories. They also do not evoke warm feelings in most humans, they are not eaten in the United States, and nobody currently thinks they are "humanlike."

However, I wanted an additional animal that was likewise a subject of experimentation but would be more controversial. I selected nonhuman primate. The British Academy of Medical Sciences has recently argued that "transplantation of sufficient human-derived neural cells into an NHP [non-human primate] as to make it possible . . . that there could be substantial functional modification of the NHP brain, such as to engender 'human-like' behaviour" should not be allowed.[17] Despite what the British Academy writes, nonhuman primates will clearly be the next laboratory subjects in this field.[18]

To avoid using the technical "nonhuman primate" in a general survey, I used "monkey." Though "monkey" is not a very precise term, referring to a broad group of species, the general public would not necessarily recognize terms like "macaque" or "marmoset."

The vignette continued with a four-level factor:

[genetically modified to have an immune system like a human / modified by having brain tissue from a human implanted in them / genetically modified to have brain tissue like a human / genetically modified to have a body that moves more like a human]

This factor describes the change to the animal. The first level of the factor is a humanized immune system, which has already been achieved by scientists, at least in mice. Because it is a nonneural change, this is the comparison case for the analysis. The next level is human brain tissue *implanted* from a human. This creates a *neuro*chimera. The claim from the previous chapter is that using tissue from an existing human entity will heighten the sense of a foundational distinction being violated. The third level is brain tissue *like* a human installed genetically. This is subtly different from the previous level in that the human brain is not *from* a human, and is thus not *mixing* an existing person.

The final level, movement resembling a human, is meant to evaluate whether reaction to a NC is truly mostly about consciousness and cognitive abilities, or whether the public's anthropology that is concerned with acting like a human has an impact. This should generate a stronger sense that the foundational distinction between humans and animals is being violated.

The vignette continues:

These are called [humanized (mice/monkeys) / chimeras / Cebirs].

The final factor attempts to evaluate the perspective described in the last chapter, in which people will perceive the foundational distinction being violated if the entity is labeled as part human, regardless of its qualities. The label "humanized" marks the entity as part human, and is a term scientists currently use to describe chimeras. Neither "chimera" nor "Cebir" references the human. I include "chimeras" to see if the term retains the frightening connotation implied by mythology; "Cebir" is a made-up term with no positive or negative connotations.

The vignette finishes with the statement:

> Scientists are hopeful that this research will eventually lead to cures for human diseases.

This is meant to emphasize the moral fact that these NCs are not being created to satisfy our curiosity but to relieve human suffering. To understand what the vignette looked like, one of the randomly selected 48 (2×2×4×3) possible vignettes presented to the respondent (without the underlining), was:

> Animals are experimented on in medical research to develop cures for human disease. Scientists plan on creating a <u>new subspecies</u> of <u>monkey</u> modified by having <u>brain tissue</u> <u>from a human implanted</u> in them. These are called <u>chimeras</u>. Scientists are hopeful that this research will eventually lead to cures for human diseases.

Survey Responses to the NC Vignette

As with the HBO vignette, the first question measured the extent of participants' approval of NC research and read: "I support the creation of these [humanized (mice/monkey)/chemeras/Cebirs]." Respondents picked from a five-point "Strongly disagree"-to-"Strongly agree" scale. Averaging across all the possible vignettes, approval proved to be mixed. Overall, 12 percent strongly agreed with research, 28 percent somewhat agreed, 27 percent neither agreed nor disagreed, 18 percent somewhat disagreed, and 15 percent strongly disagreed. If we code these responses from 1 to 5, the mean is 3, suggesting an almost perfect divide in the public's opinion. While bearing in mind that survey results are always dependent on the way a question is worded, it nevertheless appears that there is less support for NC research than there is for HBO research.

We can use the in-depth interview data to get a richer sense of what these survey responses mean. Unlike in the survey, I asked the in-depth interviewees about multiple levels of each factor in the vignette, allowing respondents to focus on comparing the different features engineered into the NC.

Again, nearly everyone turned out to be conflicted about this research. Even those who ended up in the "Strongly disagree" and "Strongly agree" categories managed to come up with potential arguments on both sides. To understand the dilemma, consider the following extensive quotation from someone thinking out loud about NCs. Danielle has a fairly explicit moral theory of the value of chimeras that is actually quite similar to the median opinion within the public-policy bioethical debate. When asked about the NC, she said:

Oh, yeah, well, I'll think out loud for a bit, which I need to do, because I don't know what I think, yet. I would say that having more human qualities does not inherently spook me. I could imagine creating a monkey with human eyeballs, so that we can learn more about blindness. And that's definitely strange, but it doesn't freak me out on a, on a deep moral level. So, it's hard for me to form an opinion for, I would say, two main reasons. One is that my opinions about animal research just in general are still kind of hazy. You know, basically, like, I hate the idea of experimenting on monkeys at all, but I'm not necessarily against it all the time either. So, you know I'm thinking about, you know, weighing the benefits and the costs. . . . As we sort of move up the scale of animal intelligence and, like, emotional complexity, I get more and more uncomfortable with it. So, I do think there's a difference between experimenting on a monkey and experimenting on a mouse. Although I could not tell you where exactly the lines are and how this all breaks down. And then, if I'm imagining monkeys with added humanlike cognitive and emotional qualities, that gets even hairier for me.

So, what's sort of funny about all of this is, I hate the idea of supporting the creation of humanized monkeys. But I also feel like that might just be sort of like intellectual weakness. I want to be able to say "No humanized monkeys" and then have other people do the, do the dirty work and then, like, save lives. So I'm really torn. I would say, if these monkeys are going to have human qualities that go beyond the physical, . . . if we're talking about human ligaments and human eyeballs and human livers, I would say, I have to go with a "Neither agree or disagree." For now. If we're thinking about monkeys with human, like, enhanced intelligence, enhanced emotional capacity—once we get into the, to the cognitive, I would say I'm "Disagree."

To get a sense of the challenge people face when thinking about this, note that the explanation above took four minutes to say (I have edited out the long pauses and repeated words).

Consider Lydia, who was asked about the NC described as a mouse with human brain tissue. Having recently discussed her approval of HBO research, she stated for the NC vignette, "I feel like I'm contradicting myself a lot, but I think that I disagree with this." When asked why, she said, "I feel like it is messing with things too much. I don't know, maybe I don't disagree with it." I then said, "Let's say the mouse gets enough brain tissue for it to be humanlike, as you put it, and they have a humanlike brain." "I would disagree with that" was her response. Eventually she summarized her position: "I'm more on the negative. I think it might be messing with things too much. Like it would be cool. I think, but I don't maybe think something being cool isn't necessarily right. I don't know, this is a hard, hard question." Note that Lydia is not the kind of inarticulate person frequently encountered in in-depth interview studies, as she was able to talk about other issues quite well. We can expect someone like this to be on the more negative end of the "Do you approve of NC research?" scale.

There are a wide variety of reasons why people are opposed to NCs. On one end is Karina, who is basically against designing new animals. She "strongly disagree[s]" with NC research because she doesn't "really like the idea of us making something to test on. . . . And I'm not saying that we shouldn't have ever done any animal testing, but I feel like designing animals just to test them is wrong."

Vanessa was concerned more about how changing the potential humanlike qualities would be offensive in and of itself, rather than about anything that would happen to the NC. She was opposed to creating a NC that was "better" and no longer really a mouse. I later asked if it was better to only change the immune system, and she said:

Why do I feel like that's different? I guess it's because you're not really messing with, like, the shape of the animal and you're not trying to change its, like, defining characteristics and you're not trying to, like, create something better. It's more that you are . . . you're changing the immune system, you're more just testing its vulnerability to different diseases than you are trying to change, like, its nature in some way. So I feel like that is different, whether that's right or not. I feel like I would be more on that side of it, yes.

She later concluded, "I feel like I am not into mutating these animals beyond, like, their natural . . . characteristics and their . . . like, natural tendencies. Changing its immune system wouldn't, like, give it humanoid characteristics or anything, but making it more humanlike in appearance or changing its brain tissue would."

As with HBOs, there were some respondents who see human health as the only real issue, even to the point of not coming up with any other moral considerations. For example, when I asked Peter his reaction to the vignette, he said, "I guess I will strongly agree." Why?, I asked. He said:

> I support the creation of these humanized mice. I want to say I agree or strongly agree. Because I don't know, I mean, mice have always been tested on. I don't like putting our man-made morals in the way of improving our way of life at this point. Seems silly, I guess, if we have the ability to make things better. I guess we should move in that direction. So yeah, I would still say strongly agree.

It often becomes evident in the in-depth interviews, as expected, that people are implicitly concerned with the foundational distinction between humans and animals even if they do not explicitly discuss it. Cultural theory would suggest that, in the case of a deeply assumed cultural distinction (such as that between humans and animals), such distinctions are so deep-seated that they are hardly even recognized.[19] It would be like asking the fish to describe the water. And indeed, these foundational distinctions often dissolve when subjected to ongoing discussion. Paul is nearly unique among the respondents because he recognizes the human-animal distinction in his own thinking. Upon hearing a description of NCs, he said:

> Yeah, this sounds a little bit more suspect to me. It sounds like— what is that movie?—like *Planet of the Apes,* or something like that. . . . I would say, strongly disagree. I don't support it. I think this is kind of taking a little bit of a bridge too far. I think that honestly, for me, it emanates more from a sense of wanting to keep my own superiority as a human and not grant that superiority to others. And I think that comes from a very, like, subconscious kind of idea of human beings wanting to dominate rather than spread that domination around.

Conclusion

We now have a baseline of public reaction to HBOs and NCs. In general, there is strong approval of HBOs but very mixed opinions about NCs. I have used the in-depth interview data to get a sense of the thought process behind these stock answers. The varying levels of support are interesting, but even more interesting is *why* there are differential levels of support, and which groups of Americans are more or less approving. In this chapter I have introduced the experimental vignettes that I will use to answer some of these questions. We'll turn to that analysis in the next chapter.

4

Consciousness, the Human-Animal Foundational Distinction, and Ephemeral Connections to Humans

In Chapter 2 I described a range of humanistic and social-science scholarship that suggested that the public is likely to have a different view of HBOs and NCs than will those engaged in the public-policy bioethical debate. Most notably, this literature suggests that the possibility that either NCs or HBOs could obtain consciousness is probably not very important.

What would be more important is, first, the foundational distinction between human and animal. While scientists and bioethicists are unlikely to believe in this distinction, it is likely that most of the public does, and that this would make the public less supportive of NC research.

The second major factor is the belief that humans retain ephemeral connections to their disembodied parts. For someone who believes in these connections, HBOs violate the human-object distinction, because the object (the HBO) is still somewhat human. A human is then being treated like an object. Scientists and bioethicists, by contrast, are unlikely to believe in these connections and are instead likely to see the body as purely material. Belief that humans retain these connections to their disembodied parts may decrease support for HBO research. I suspect that the public's "yuck" reaction when told about these technologies results from breaching these two foundational distinctions.

Disembodied Brains. John H. Evans, Oxford University Press. © Oxford University Press 2024.
DOI: 10.1093/oso/9780197750704.003.0004

The previous chapter described the research instruments I created to look for evidence of the public's views. Let me offer a few more words about such surveys. If I find a 15-percent difference between any two sub-populations in their agreement with something like HBO research, that is a fairly substantial difference. An effect larger than 15 percent would probably already have been commonly evident without having to conduct a study. It is usually interesting that there is a difference at all, because this tells us something about the experiences of those sub-populations.

Consciousness

The survey provides evidence regarding whether the public is concerned about consciousness. In the HBO vignette component of the survey, the respondents were randomly divided into four equal groups each seeing one of four different vignettes; one quarter each saw a vignette in which (1) the HBO would not achieve consciousness, (2) the HBO could achieve the consciousness of an insect, (3) the HBO could achieve the consciousness of a pig, and (4) the HBO could achieve the consciousness of a human fetus. If the level of consciousness is the determinant of support or opposition to HBO research, then we would expect increasing opposition in the order I describe for the factor levels above.

The formal results of the statistical analysis are reported in the endnotes to this chapter and in tables available online. The analysis shows no difference in level of opposition between those who considered HBOs with no consciousness, insect-level consciousness, pig-level consciousness, and human-fetus-level consciousness.[1] The finding regarding the human fetus is particularly surprising, in that the vignette was meant to evoke the idea of a conscious human trapped in a dish,[2] reflecting the tone of media stories surrounding the discovery that HBOs have the brain waves of human fetuses.

Perhaps an effect of level of consciousness would have been found if I had added the option of "10-year-old human," but it did not seem credible that a ½-inch piece of tissue without any experiences could obtain that level of consciousness. That said, comparing responses to "no consciousness" and "consciousness of a human fetus" is a fair test of the idea that the public is concerned with levels of consciousness. It appears that the public is not concerned. The vignette also contained a factor describing awareness in two ways, one more humanlike than the other. These results also fail to support the consciousness theory.[3]

Let me put a finer point on this. The vignette shows that the public does not use levels of consciousness on their own when examining this topic. When they do not realize that the person asking the survey questions is comparing levels of consciousness, consciousness is not "what comes to mind." But if a bioethicist were to argue to the public that levels of consciousness were really important, I suspect they would agree. Another way to put this is: Does the public think consciousness is important, or could they come to agree with the bioethicists' claims that consciousness is important? I think the answer is, the public does not think it is important, but could come to accept the bioethicists' claims. I believe that policy should follow the public's actual views, to the extent possible, instead of the views of specialists to which the public might eventually agree.

"What comes to mind" and "agree to" require different survey techniques. To discover the former, I can use a vignette in which the respondents do not know that I am looking to see if they use a continuum of consciousness to evaluate these technologies, since each of them is asked to consider only one level of consciousness (e.g., insect). For the "agree to" question, we could use a standard survey in which they would see a question explicitly asking about the importance of consciousness. Asked such a question, respondents would realize that differential consciousness is important to the researcher and would indicate their level of agreement.

A standard survey item included later in the survey, after the vignette, explicitly suggested to the respondents that consciousness is important, essentially asking for their evaluation of the bioethicists' moral concern: "If human brain organoids obtain higher levels of consciousness, we should treat them more like we treat born humans." In response, 13 percent of respondents strongly disagreed, 16 percent somewhat disagreed, 35 percent took the neutral position, 26 percent somewhat agreed, and 11 percent strongly agreed. So perhaps half of the public *does* care about consciousness at an abstract level—just not at the level that bioethicists care about it (as bioethicists would join the 11 percent of the public who strongly agree with this statement). Despite not using consciousness in their evaluation, a good portion of the public could be convinced to do so.

This difference between an experimental vignette and standard survey questions can be shown with the in-depth interview data. In these interviews respondents were shown versions of each vignette instead of just one, and they could then calibrate their responses. Respondents discussed the differences between the vignettes.

Some interview respondents did not distinguish between these levels of consciousness, basically saying that we kill insects, pigs, and human fetuses all the time, so none of that is problematic. For example, Marian, in responding to the HBO vignette, said she was comfortable with destroying the HBO with the possible consciousness of an insect. How about the self-awareness of a pig? I asked. She responded, "We slaughter pigs for food, you know. We do kill things that are self-aware and try to do it responsibly, but also there are purposes that animals serve." If the vignette had specified the self-awareness of a human fetus, she said her answer would not change, and instead argued about how she would know about its consciousness anyway: "How would we know that something was that self-aware? And if, in the process of discovering that these brain organoids are that self-aware, would we also simultaneously be discovering that a lot of things in our environment are far more self-aware than we suppose?"

However, when I walked other respondents through their reactions to the different possible levels of the consciousness factor in the HBO vignette, most of them alluded to a consciousness continuum. For example, Bill, who we met in the last chapter, was talking about whether it would be OK to destroy an HBO. To test the levels of the vignette, I changed my description of consciousness. He had started with the "will not develop consciousness" level and said of HBO destruction, "I do not know why I agree. That sounds agreeable. Interesting. I think I agree with that."

I then asked, What if the HBO had awareness of its surroundings, to the level of an insect? He responded, "I've always kind of viewed this as stuff that is non-aware. . . . [But] when something begins to interact with its environment enough that it's aware of who it is, or that it's feeling pain, it wants to change the scenario, [then] I'm a little less comfortable with it."

How about the level of a pig? His response: "Yeah, ya know, I think we're reaching a good endpoint." I asked what was different between the insect and pig cases. He responded, "Their reasoning and interaction is at the animal level. To me there definitely seems to be a lot more awareness of even abstract things. You know the animal sees. The hunter steps behind the tree, they still know there's a hunter there, you know, even though the tree prevents them from seeing it. I don't know if insects are at that level." And, finally, I asked about consciousness like that of a human fetus, which was the underlying issue in the media reports on the brain-wave experiments. How does this affect your comfort level? I asked. His response: "Uncomfortable."

The respondents were aware from the way I asked these questions that levels of consciousness were supposed to be important to their answer. They then invoked a basic idea of levels of consciousness whereby a pig is more important than an insect. But when respondents did not have the ability to compare, as in the experimental vignette version for which they were randomly assigned to only see one of these consciousness levels, the public's response was that their approval is not contingent on the consciousness of the HBO. Though the public is not currently concerned about consciousness, the public could be convinced to be concerned about consciousness.

The Human-Animal Foundational Distinction

To what extent does the public believe in a foundational distinction between humans and animals? I have a range of analytic puzzle pieces that, when combined, show a coherent picture. I start with two fairly direct survey questions that were asked toward the end of the survey, well after the HBO and NC vignettes.[4] These are not part of an experimental vignette, so they measure the extent to which the respondents agree with the ideas presented to them. Importantly, these questions did not ask about HBOs or NCs but were instead very general; the in-depth interviews show that people interpreted them in the most general of ways. I asked respondents whether they agreed that, "in general, humans are more similar to animals than they are different," and most respondents agreed. While only 6 percent strongly disagreed and 13 percent somewhat disagreed, 21 percent took the neutral position, 43 percent somewhat agreed, and 18 percent strongly agreed.

In the in-depth interviews, it was typical for respondents to interpret the question in the most general of ways. For example, Brian said to me: "We coexist in the same ecosphere . . . with slight variations. Obviously, some are underground or underwater . . . but we're all carbon-based life-forms [using] oxygen primarily." Lydia simply said, "We're all just trying to eat and sleep and make babies." I think that those who agree that we are similar at this level of generality think (to use the metaphor from earlier chapters) that humans and animals are on the same continuum. However, this question cannot reveal whether they think there is a large gap in the continuum between humans and other animals.

Those who saw a difference were thinking more specifically. Bill said that he disagrees that humans and animals are similar, saying "Again, I think it's the human consciousness, to decide things. We're all going to eat, we're all going to go to sleep, we're all going to look for shelter, we're all going to look for companionship and all the things that the animals do too. But I think we go about it one more deeper level than they do."

Sophia also said humans and animals are different. Why? I asked. She said, "The difference lies with the soul and with the power of the brain in humans vs. animals, in the fact that we can plan, we can act differently from animals, and we should be acting differently from animals. We have a conscience. Ethics and whatnot makes a big difference."

The survey also asked for the participant's level of agreement with the statement "A chimpanzee with a human brain would be human." By assuming no gap in a human-animal continuum, this is a direct measure of the most concrete version of violating the foundational distinction. Respondents had the opposite reaction to this, with 29.2 percent strongly disagreeing, 26.3 percent somewhat disagreeing, 25.6 percent taking the neutral position, and only 13.0 percent somewhat agreeing and 5.9 percent strongly agreeing. Obviously, this question represents an extreme statement, and there could be all sorts of reasons to disagree with it. That said, when the question is framed at this level of specificity, very few people see no distinction between humans and animals.

The more important question is whether those who most strongly believe in the human-animal distinction also are opposed to NC research, which would suggest that belief in the divide determines one's views of NCs. Looking at the general agreement that NC research should continue question asked after reading the vignette, we see that those who believe in a stronger distinction between humans and animals are more opposed to NC research. While it is difficult to describe how strong an effect is in a survey, I use a method that relies upon describing the difference between exemplar respondents.[5]

Of the respondents who *least* agreed that humans and animals are similar, 30 percent somewhat agreed or strongly agreed that NC research should continue, while those who *most* agreed that humans and animals are similar, 45 percent somewhat or strongly agreed.[6] This is a fairly large effect by survey standards. Of the respondents who *least* agreed that the chimpanzee with a human brain would be human, 36 percent somewhat or strongly agree that NC research should continue, while of those who *most* agreed that the chimp would be human, 46 percent somewhat or strongly agreed. This is a more moderate effect. Combined, these analyses support the idea that opposition is driven by belief in a foundational distinction between humans and animals.

The NC Vignette and Approval of NC Research

The NC vignette was designed primarily to evaluate the theory that people will oppose creating NCs because NCs violate the foundational distinction between humans and animals. The logic is that the NC will be considered an abomination and therefore should not be created. The analysis shows that those who saw the vignette in which the animal involved was a monkey (instead of a mouse) were less likely to approve of chimeras, with a 13-percent difference in selecting "Agree" or "Strongly agree with continued NC research."[7] All would agree that monkeys are cognitively much superior to mice, so if the public were concerned with levels of consciousness, we would presumably see much more than a 13-percent difference here. Instead, what we see, especially when combined with the puzzle pieces I will show below, is that the public is primarily concerned that the foundational distinction is being crossed with any animal, whether mouse or monkey.

The vignette also had four variants involving possible modifications to the animal that the respondents were randomly assigned to see: (1) a human immune system, (2) implanted human brain tissue, (3) a brain like a human, and (4) movement more like a human. The analysis shows that respondents who saw either of the two "brain" scenarios were more opposed to NC research than those who saw the "immune system" scenario.[8]

The foundational distinction is threatened when core elements of the animal and the human are mixed. Since the brain is where the public thinks capacities such as "ability to love" are housed, such modification therefore threatens the human-animal distinction. However, one could also argue that the two brain scenarios indicate to the respondent that the resulting animal, now vested with some human brain tissue, would have greater consciousness. So perhaps that result is ambiguous in adjudicating between the consciousness and human-animal distinction perspectives.[9]

However, the 25 percent of respondents who saw the change as making the animal "move like a human" were more opposed to NC research, which suggests that it is not consciousness that is registering with people but qualities that are part of the definition of the human used by the public. Human movement violates the human-animal distinction because, as discussed in Chapter 2, people define humans as those entities that look or act like humans. However, this effect is relatively small.[10]

Finally, the respondents' reaction to the labels applied to the chimera supports the idea that the public is concerned about the human-animal distinction. In Chapter 2 I theorized that one way to violate the human-animal distinction is through labeling, which could invoke the idea that the entity is part-human. In the vignette, respondents were randomly assigned to see one of three labels for the NC: "humanized mouse"/"humanized monkey," "chimera," and

"Cebir."[11] For this analysis, the label that other labels are compared to is "humanized," which is currently used by scientists. The results show that it is actually "humanized," as compared to "chimera" and the made-up label "Cebir," that generates disagreement with creating the entity. I interpret this to mean that reminding the respondent that the NC is a mix of human and animal triggers the violation of the human-animal distinction, and thus disapproval of this area of research.

As an aside, this analysis allows for a contribution to a separate and smaller pragmatic debate among scientists. A prominent report on NCs and HBOs, focusing on the proper label for these entities, avoided the term "chimera," partly because scientists do not use the term but also because it is suspected of generating unjustified fear in the public due to the connection with mythological monsters.[12] My own results suggest that "humanized" is scarier than "chimera," probably because most respondents wouldn't know the mythical origins of the term and so would likely assume that it's a made-up name like "Cebir."

Is There a Foundational Distinction between the NC and the Human?

The previous analysis assumed that violating the foundational distinction is indicated by disapproval of the research. This issue was directly broached when I presented respondents the following statement: "There is not really too much of a difference between these [humanized (mice/monkeys)/chimeras/Cebirs] and humans." As you would expect, given the earlier questions about differences between humans and chimpanzees, only 4 percent of respondents overall strongly agreed with this assertion. An additional 14 percent somewhat agreed, 27 percent neither agreed nor disagreed, 25 percent somewhat disagreed, and 30 percent strongly disagreed.[13]

If it is capacities like consciousness that make people think a human and animal are similar, then the NC with a human brain should be considered more like a human than the NC with the human immune system. However, seeing the "brain tissue like a human" variant did not in fact lead to more belief in similarity than the "human immune system" variant. In comparison to "brain tissue *like* a human," "brain tissue *from* a human" is designed to more directly invoke a violation of the foundational distinction because in the latter scenario the tissue comes from an actual existing human. However, the analysis shows that the respondents who saw "brain tissue from a human" are actually more likely to think NCs are *different* from humans. This response seems to provide support for the foundational distinction theory.

Similarly, seeing the modification in the vignette as installing "human movement" also leads to seeing *more* difference between the NC and humans. Though we might think that human movement would increase one's perception of humanness, we actually see the inverse. Similarly, and as before, labeling the NC as "humanized" (rather than" Cebir") actually leads to respondents seeing *more* difference. (There is no difference between seeing "chimera" compared to "humanized.") As the foundational distinction is violated, I think we are seeing the disgust response at work. On seeing "human brain"—particularly brain tissue *from* a human—and the label "humanized," respondents say to themselves, "No—that does *not* make them like a human!"

Consciousness, the Human-Animal Distinction, and Treatment

Another set of questions sought to discover respondents' views of how we should *treat* HBOs and NCs. These questions served two purposes: First, they showed the moral status of these entities for the public, which is part of their anthropologies. Second, they indicated whether the public was concerned with consciousness or the human-animal foundational distinction.

After the general question about whether to continue HBO research, the survey asked for responses to the statement "I am comfortable with these brain organoids being destroyed when the experiments are done." The public evidently does not think such entities have high moral status: only 4 percent selected "Strongly disagree" and only 9 percent selected "Somewhat disagree." That said, HBOs were obviously granted more respect than a piece of fingernail paring; 31 percent selected the "Neutral" position, while 28 percent chose "Somewhat agree" and an identical 28 percent chose "Strongly agree." Clearly, while the overall public does not think of HBOs as having anything close to the status of humans, that does not mean that they have no status at all.

As you would expect, another analysis shows that those who think it is fine to destroy HBOs are the most supportive of the research in the first place, and those who are most opposed to destroying them are most opposed to this research—that is, people who think HBOs have some moral status do not want to create them.[14]

The consciousness theory would predict that the respondents who saw vignette variants where the HBO had higher levels of consciousness (insect, pig, human fetus) should be increasingly reluctant to destroy the HBOs, because these HBOs would have more moral status. But analyses show no difference between those who saw these higher levels of consciousness and those who saw that HBOs are expected to not be conscious (the comparison group).[15] Levels of consciousness do not seem to be important for how we treat HBOs.

We can also look at the treatment of NCs. In the public-policy bioethical debate, the consensus is that, the more human capacities a NC has, the better it should be treated. Does adding human brain tissue lead the public to think we should treat the animal better? After presenting the NC vignette, and after asking about approval of NC research, I asked for the evaluation of the statement "We should treat these [humanized mice/monkeys//chimeras//Cebirs] better than we treat other [(mice/monkeys)]." Only 7 percent strongly agreed, 17 percent somewhat agreed, 42 percent took the neutral position, 16 percent somewhat disagreed, and 18 percent strongly disagreed. That is, a majority think that adding human qualities does not increase the status of the animal. This seems to support the human-animal distinction that would hold that an upgraded mouse is still a mouse and an upgraded monkey is still a monkey.

We can also examine the vignette factors.[16] If the public thinks like the ethicists, the respondents who saw the vignettes in which the NC had brains that were more human should think that those NCs should be treated better. That is, if it were to become "more human," we should upgrade its treatment. But the analysis actually shows that seeing either of the two human brain-tissue additions, compared to seeing the change as adding a human immune system, does not incline the respondents in favor of improved treatment.

In fact, those who saw a vignette variant in which the NC was capable of "human movement" actually were slightly *less* likely to think the NC should be treated better than other mice or monkeys. This is another piece of evidence supporting the theory that there is a disgust response to violating the human-animal distinction. In Western culture, life-forms that violate foundational distinctions have been considered "monsters." So instead of regarding the NC as more like a human and thus deserving of better treatment, the respondents apparently see the animal that moves like a human as a monster and want to treat it worse than other mice or monkeys. This is like the villagers' response to Frankenstein's distinction-violating monster—destroy it.

The influence of labeling has a similar effect. When the NC is described as "humanized," people are *less* likely to say it should be treated better than those who saw it labeled "chimera."[17] Those who saw "monkey" were less likely to think the NC should be treated better than those who saw "mouse."[18] This is consistent with the idea that the proximity of monkeys to the border of the foundational distinction makes them more of a threat. Again, these seem to trigger disgust reactions, in which the violation of the human-animal distinction leads people to conclude, "Treat the humanized entity worse."

Finally, I asked respondents to evaluate the statement, "If enough human qualities were put into the [humanized mice/monkeys// chimeras//Cebirs], we should think of them as having the same importance as born human beings." This is a distillation of the moral perspective of bioethicists into a phrase. Only 7 percent selected "Strongly agree," and 18 percent "Somewhat agree"; 28 percent selected the neutral response, 20 percent somewhat disagreed, and 27 percent strongly disagreed. That is, very few members of the public agree with the extreme version of bioethicists' reasoning that species is morally irrelevant and that all that matters is capacities. (In fairness, probably few people could imagine a NC having the same human qualities as an actual human.)

If capacities were driving the moral status of the NC, the NC with more human traits (e.g., a more human brain) should be more likely to be equated with humans, but it turns out that this is not the case. The different qualities that people see as installed in the NC do not affect their responses.[19] The effect of labeling shows the opposite of what the capacities theory would predict and supports the idea that there is a disgust-driven backlash effect. Respondents who saw the "humanized" label in the vignette were less likely to think the NC should have the same importance as humans compared to those who saw the "Cebir" label.[20]

So far, I have shown analyses, metaphorically described as puzzle pieces, suggesting that, when all is put on the table, the public is not very concerned about the consciousness of HBOs and NCs. Rather, the pieces suggest that the concern is with the human-animal distinction. The strongest pieces of the puzzle even suggest that there is a backlash of disgust, and that suggesting to the respondent that an entity is a mix of human and animal on key components of humanity results in a desire to treat it worse. This effect is akin to how an average member of the public would want to treat other violators of foundational distinctions, like zombies.

Ephemeral Connections to Actual Humans

In Chapter 2 I relied upon existing scholarship to suggest that the definition of a human used by the public leads to thinking that extra-bodily human tissue retains a connection to the actual human. This in turn would lead to a breakdown of the human-object foundational distinction, because the object (the HBO) has a part of an actual human in it or "is" a human in an ephemeral sense.

In the medical perspective, disembodied human tissue has no particular status and is usually discarded and burned. The public-policy bioethical debate discusses how people own their tissue, and agrees that it cannot be used without permission. However, for the general public, the relationship between actual humans and their disembodied tissue probably runs much deeper and will impact how people view HBO and NC.

I was also able to conduct analyses to evaluate these hypotheses. After the HBO vignette, and after the question about whether this research should continue, I asked respondents to evaluate the statement "Human brain organoids should be thought of as an extension of the person who donated the skin cells used to make it." Only 13 percent strongly disagreed, 15 percent somewhat disagreed, and 32 percent took the neutral position; more interestingly, 28 percent somewhat agreed and 13 percent strongly agreed. Skeptics may suspect that these respondents were reading the survey too fast, but there is no evidence to suggest this.[21]

Interviewees' Understanding of Ephemeral Connections to Actual Humans

An examination of the in-depth interviews shows that there are three interpretations of this question—that "extension" may be legal, biological, or spiritual. First, a few mentioned the idea that people own their cells and that scientists cannot do anything to those cells without the donor's permission, which is a critical issue in the public-policy bioethical debate. For example, John says, "It comes down to consent. . . . You would have signed away that consent, otherwise you wouldn't have been allowed into whatever the program happens to be."

A group of respondents strongly disagree with "extension" because they take a materialist perspective. For example, Karina says, "If my finger was amputated at that point, it's not an extension of me anymore—like, it's a detached dead limb." Similarly, Paul says, "I would say, I strongly disagree, I mean, when somebody . . . donated a kidney and that's implanted in somebody else, that's no longer part of you, you know."

Respondents who see a connection with the donated body part are usually referencing a genetic connection between the donor's cells and the HBO. This may be the result of decades of public discourse that emphasizes not only that we "are" our DNA, but also that our particular DNA is unique to us.[22] Steve saw a general genetic connection:

> Well, because the person who made the donation has some type of connection, you know, to the organoid that was created. Yeah, I mean, it's not going to have a personality, or anything like that. But . . . skin cells . . . are specific to an individual, which could also be, you know, looked at, as they say, if that person that donated the skin cells had a particular disease or something that we were looking at as far as research concerned, it could be, you know, connected in that way.

One step stronger than just "it has the same DNA" are analogies to children. Robert says it is "similar to a person being born. So, if the brain organoid was an extension of the person who donated the skin cells, that would be similar to that same person giving birth." Dan says, "It's like a child being an extension of their parents. A child is not their own person from, like, the minute they're born. Oh, the parents should be held responsible for them. But, yeah, it's like an extension." That people make a family connection suggests a stronger link and potentially higher status for the HBO.

The idea of extension was often rejected because the HBO was small and limited, but if what was grown was more substantial, then "extension" would be justified. Isabel said, "Although I don't actually think it's going to be an extension of a person, but it is like . . . replicating a human, to a certain extent, whereas this is like a decent body part, and that just to me does not feel as . . . replicable." Similarly, Robin said, "I think I would disagree with that as well. Because I think a skin cell and an entire human being . . . Like, it's still . . . a piece of science in a lab—it's not like a human being in the streets. Although I think it would be nice to acknowledge that it came from someone, I don't think that it's necessary you treat it, like, as an entire extension of that person." Note the implication here, which is that if the HBO was more representative of an entire human, it would be more of an extension of the donor.

Finally, there are respondents who see an HBO as an extension because of a more diffuse spiritual connection. Eric said, "I think there's something about our beings that make up everything about who we are. And that is important always to acknowledge what that is, no matter what lens we look at that through. It should always be traced back to that. I think that that helps it ensure that it lasts forever. . . . I feel like it needs to be, in that way, attached."

The Impact of Belief in Extension on Treatment

The survey showed that more people agreed than disagreed that an HBO was an extension of the human who provided the cells from which it was grown. Does seeing a connection to an actual human mean we should treat an HBO more like a human?

The analysis shows that those who see the HBO as an extension are much less likely to agree that HBOs can be destroyed. Of those who most agreed that an HBO is an extension, 48 percent also somewhat or strongly agreed that HBOs could be destroyed.[23] Of those who least agreed that an HBO is an extension, 73 percent also somewhat or strongly agreed that HBOs could be destroyed. Thinking of HBOs as an extension is also strongly associated with thinking that an HBO with human traits should be treated more like a human. Of those who agreed that HBOs are an extension, 68 percent somewhat or strongly agreed that HBOs should be treated more like a human. Of those who did not agree that HBOs are an extension, 4 percent somewhat or strongly agreed that HBOs should be treated more like a human.[24] Those who see ephemeral connections plainly want better treatment of HBOs.

HBOs Having Shared Thoughts with the Donor

Some of the interpretations of the extension question—for example, that extension means "ownership"—are consistent with a materialist understanding of the human used in the public-policy bioethical debate. I also evaluated a strong ephemeral connection that clearly does not assume materialism, which is the transmission of thought, similar to what we see in studies of organ transplantation. I asked the respondent to evaluate the statement, "If a human brain organoid could think, it might have the thoughts of the person who donated the skin cells." While this is outlandish from a scientific perspective, only 12 percent of the public strongly disagreed, 15 percent somewhat disagreed, 35 percent took the neutral position, 29 percent somewhat agreed, and 9 percent strongly agreed. Those who somewhat or strongly agreed thus outnumbered those who somewhat or strongly disagreed.

Though we can acknowledge that most Americans do not understand the scientific views of human biology, that is actually beside the point, since they are nevertheless part of the public whose collective views should determine science policy. A good portion of the public believe humans possess both body and soul, and thus are likely to believe in the possibility of the most ephemeral connections, such as shared thoughts.

Interviewees' Understanding of Shared Thoughts with HBO

There were a range of responses to the interview statement about an organoid possibly retaining the thoughts of its donor. The more scientific responses will be quite legible to most readers. A good number of respondents basically said, "That does not make sense," or "How could that be?" The more articulate explained that thoughts are not contained in each cell, so how could they be the same? As Peter said, "My brain only knows what I've experienced, so I don't see how it could have the thoughts . . . I can't even imagine." John said, "It kind of seems like a non sequitur because you know consciousness is not just cells, it's also experience. So obviously even if it was a clone version . . . it would not be the same."

A very large group of people are utterly unsure what to make of the claim and therefore pick a more neutral response. Vanessa selected "Disagree" (not "Strongly disagree") and clearly was inclined to agree with those in the first group but was not entirely sure: "I literally have no idea. . . . I haven't researched that much into where, like, personality and things come from, whether it's, like, nature versus nurture, you know. If you're, like, genetically, or if you . . . Or if it's what you're surrounded by when you grow up, but I lean a little more towards like . . . you grow into your personality and who you are." Brian says, "I don't know enough about stem cells, . . . so I would neither agree or disagree. . . . It seems really unlikely." Farrell says, "I don't quite see how that would make sense, but I guess . . . I mean, maybe. I don't agree, nor do I disagree. I could see arguments for both sides."

A contingent of respondents thought that DNA in cells could transmit thoughts. Richard says, "It might. . . . But as the same DNA, I mean, yeah, probably agree with that." Along with those who had heard that trees communicate with each other, and thus anything similar was possible, there were others who also saw more diffuse connections. A few were directly aware of reports of memories being transmitted via organ donation. Ross, when asked if an HBO "might have the thoughts of the person who donated the skin cells," said:

> I kind of agree, because in some of the books I have read you see stories. There was one, a few years back, where a football player died . . . and the parents donated the heart to an older woman in New York or something. When she woke up she had this strange craving for banana Popsicles, and that happened to be his favorite treat. . . . I have read and seen too much stuff for donating organs and people got the memories of the previous owners. So, yeah, something like that can be transferred to the new owner.

Similarly, Paul was using an example of a heart transplant, which prompted us to ask whose heart it was. He answered, "I know it was somebody else's heart, but now it is currently the current owner's heart. But when I think how little we know about . . . these things it is insane. . . . There might be some kind of influences. . . . If I have a soul, . . . and I think I do, that has so much to do with electrical impulses and signals." He continued with an example of how a transplanted organ could be directing the activity of the recipient: "So say that I plant trees, all in a certain area, and maybe I keep coming back to this area for the rest of forever. I've drawn this time and space, this energy, this plant—it's still here after I'm gone. . . . So I think there are these remnants of things that can be leftovers from the spirit before, the soul before."

To scientifically informed readers, this sounds preposterous. However, for those who believe in spirit or soul, the idea of nonmaterial connections between living entities was not so unlikely. In fact it was plausible enough to push their response a step up the scale from, for example, "Somewhat disagree" to "Neither disagree nor agree" as well as to generate generalized disgust with HBOs.

Further Evidence for Ephemeral Connections Based on Belief in Spirit or Soul

These ephemeral connections will seem outlandish enough to many readers that I want to provide further evidence that belief in these connections comes from belief that there are spiritual forces in the world. The largest institution in the United States that teaches ephemeral connections is Christianity, and, in the United States, only Christianity is large enough for me to analyze with a national survey. Belief in ephemeral connections is very much like other beliefs found among Christians.

To also take a few obvious examples; most Christians believe that people have a nonmaterial essence called a soul that can survive the body's death. Christians generally believe they are in communication with a divine entity that is beyond matter and cannot be seen. They retain a relationship with dead people whose souls are in heaven. Christians believe that God spoke through certain historical figures, and a subset believes that God continues to speak through either representatives on earth (e.g., the Pope) or individuals possessed of the spirit (e.g., in Pentecostalism or Quakerism). To the Christian mind, the idea that HBOs are an extension of a human, or that there could be a thought connection to an HBO, is not as fundamentally weird as it would be for an atheist.

We will be delving more deeply into religion in the next chapter, but for now I looked at who in my survey sample believes in the ephemeral connections with HBOs. Adjusting the figures to account for differential education levels and other demographic and attitudinal features, we find that adherents of every active Christian group except the most liberal ones are more likely than the nonreligious to believe in these ephemeral connections.[25] This supports my interpretation that these ephemeral connections are not just biological or social but at least partly spiritual.

The Impact of Belief in Shared Thoughts on Treatment

Does this stronger version of an ephemeral connection lead to increased status for HBOs in the public's mind? Those who agree that an HBO may share thoughts with the cell donor are indeed slightly less likely to think that HBOs should be destroyed; those with the highest and lowest agreement with the "thoughts" question differ by 8 percentage points in selecting "Agree" or "Strongly agree" with the statement that HBOs can be destroyed.[26] The idea that, if an HBO were to obtain enough important human traits, we would treat it more in the way we treat humans is more strongly predicted by believing in shared thoughts. Those with the highest and lowest agreement with the "thoughts" statement differ by 34 percentage points in selecting "Agree" or "Strongly agree" that the HBOs should be treated more like a human.[27] That this effect is stronger than the previous one makes sense, in that *if* an HBO is described as more human, then the ephemeral connection to the actual human will be stronger. Only an HBO with more human traits could have thoughts.

In sum, a large portion of the public believes that humans are not just material but also have a nonmaterial spiritual component. For these people, imagining an ephemeral connection with disembodied human parts is not outlandish. And, as we have seen, belief in these ephemeral connections affects the public's view of how to treat HBOs. Theory would suggest that the public thinks that the foundational distinction between humans and objects is being violated and that these HBO objects have part of a human within them. As we have also seen, those who think we should treat HBOs better are less in favor of the research in the first place.

Conclusion

The reason NCs and HBOs are controversial is that they implicate the human, so one's evaluation of these technologies will depend at least in part on what one thinks a human is. The professional participants in the public-policy bioethical debate tend to use the academic philosophical or biological definitions of the human. As a result, the main ethical concern about NCs and HBOs turns out to be that they could achieve an unacceptable level of "humanlike" consciousness.

In this chapter I have examined the public's views, and the public has different definitions of the human. Social science rarely examines the underlying cause of the public's views on issues like these, because they are so hard to measure. There is no one test that reveals the public's views on the subject, so I have instead created, metaphorically, pieces of a picture puzzle that do go together, although not precisely, to paint a coherent picture of the public's views.

The public is unlikely to be primarily concerned with the consciousness of HBOs and NCs. Though they will agree when presented with the idea that consciousness is paramount, their own instincts lie elsewhere. Regarding NCs, the public is primarily concerned about violating the foundational distinction between humans and animals. Regarding HBOs, the public is primarily concerned that extra-bodily tissue retains an ephemeral connection with the original human. These findings are important for our understanding of the HBO and NC debate, but we also need to account for the fact that these technologies are instances of a larger category of biotechnologies for which segments of the public may have different views. I turn to these analyses in the next chapter.

5

Views of Nature, Religion, and the Cultural Authority of Science

We have been focused on how the public's definition of a human ultimately impacts its views of HBOs and NCs by emphasizing the foundational distinctions between humans and animals as well as the distinctions between humans and objects (HBOs). But, given the complexity of society, there is never just one explanation for public opinion on a given subject. Rather, a number of social forces combine to produce approval or disapproval of HBOs and NCs. To truly understand the public's views, we need to examine additional reasons for support and opposition that are broader than NCs or HBOs. HBOs and NCs are biotechnologies, and social groups can be expected to have positive or negative reactions to biotechnologies in general. This chapter will primarily focus on people who have differing views of the relationship of humans to nature, different religious groups and beliefs, and distinctive orientations toward science.

Demography and Support for HBOs and NCs

Most social-science analyses of public opinion, including those devoted to biotechnology, focus on demographic differences in opinion. These analyses are often informative because they map onto readily observable groups in society, such as people of different ages. For example, when social scientists measure demographic

Disembodied Brains. John H. Evans, Oxford University Press. © Oxford University Press 2024.
DOI: 10.1093/oso/9780197750704.003.0005

characteristics like female gender, what they are really saying is that women tend to have a different social experience in the United States than men and that this differential social experience leads to different views of biotechnology.

One reason for the focus on demography is that knowing a person's demographic identity usually makes it easier to imagine which social experience led to a given attitude or opinion. For example, consider education levels. We can imagine (and could demonstrate with a separate project) that those with more education are exposed to more science through classes or in the workplace. Higher education, including education in nonscience fields, also tends to require that students employ the kind of Enlightenment reason that underpins science; in a humanities classroom, for example, you cannot offer the explanation "God caused it." Thus, more education tends to produce distinctive views of HBOs and NCs. I will examine the social groups most analyzed by other scholars, but without focusing too much on demography, because the existing literature does not suggest that demography is very important for predicting views of HBOs and NCs.[1]

In studies of attitudes toward biotechnology, gender is not a consistent determinant.[2] Indeed, it is a little hard to imagine what different social experiences would lead to gender difference in views of HBOs and NCs, and my analysis of the survey shows that there are indeed no gender differences in approval or disapproval of HBO or NC research.[3]

Race is more likely to influence trust in medicine than views of basic science. For example, African Americans are less likely than white people to agree that scientists understand the Covid pandemic and that the scientists involved with Covid share their values.[4] Consistent with this expectation, I find no racial differences in view of approval of HBO or NC research.

Respondents with higher levels of educational achievement are usually shown to be more supportive of technology. But in my analysis, education level does not predict people's views of HBOs;

people with less than a high-school education have the same views as those with graduate degrees. For NCs, more education is actually associated with more opposition, which is the inverse of what is found in most studies of attitudes toward biotechnology.[5] It may be that, in other studies, the education measure has actually been a stand-in for other attributes that my own study measures directly (see below), which would explain this anomalous finding.

My own research reveals that older respondents tend to approve more of HBO and NC research.[6] Is this an age effect or a period effect?[7] If it is simply a reflection of age, in which opinion tends to be determined by one's stage of life, then we can expect the 25-year-olds to have become more supportive by the time they are 60-year-olds. Perhaps older people are more supportive because they are more fearful of Alzheimer's and other neurological diseases. If it is instead a period effect, then changes in American culture mean that the young are going to remain more opposed to the technology as they age because they were raised and educated differently than the average 60-year-old. Perhaps younger people in more recent decades have perceived that human intervention into the natural world is suspect, not a point of pride. But this is only speculation; my present research design cannot establish whether this approval is an age or a period effect.

Orientation to Nature

Biotechnology modifies the existing world. Therefore, one's orientation toward nature is likely to determine one's views of biotechnology in general, including HBOs and NCs. Environmental ethicists make a distinction between conservationism and preservationism. Conservationists are in favor of the wise use of the natural world for human purposes; think here of cutting trees for human use while planting new trees in their place. (Conservationism in public discourse has a more general meaning.)

In contrast, preservationists see nature as more sacred and not to be disturbed by human actions; think here of Henry David Thoreau's focus on untrammeled wilderness untouched by humans, or Aldo Leopold's view that "untouched nature is exactly how nature should be. Millions of years of biological history have endowed it with a moral or even religious significance."[8] The preservationist view is what we often mean by "natural," which John Stuart Mill famously defined as that which is free of human interference.[9]

This basic distinction, which has been described in various ways, represents the core debate in justifications for environmentalism. To take a concrete example, a National Academies of Science report about genetically transforming animals asks what human values are at stake, and one such value is the potential impact of modified organisms on the environment. But this has both conservationist and preservationist interpretations; that is, impact can be "understood both in terms of outcomes for people and, for some individuals and cultures, as a concern about the environment in its own right."[10]

The same distinction is often labeled with different terms. The National Academies report goes on to describe this distinction as between "intrinsic" and "anthropocentric" values, which gives us more insight into the divide. "Anthropocentric" would mean we are focused on "functions that are vital to humans, communities and societies," whereas "intrinsic" would mean we must evaluate "environmental outcomes not only in terms of outcomes for humans but also in terms of their effects on the environment itself." For example, we might support saving endangered species because of "their economic or medical usefulness"; alternatively, "they may also be considered valuable in and of themselves."[11] Philosopher Christopher Preston puts it slightly differently, acknowledging both those in technological debates who want to "diminish our footprint on the planet" and those who want to shape nature "confidentially, deliberatively and sometimes ruthlessly, all according to the best abilities of our technical experts."[12]

According to Preston, if we continue on our path of genetic and other modifications of nature, "humans stand on the verge of turning a world that is found into a world that is made." In the synthetic age, "the world is thoroughly reconstructed, from the ground up, by molecular biologists and engineers, marking the beginning of the planet's first Synthetic Age."[13] As another critic of technology says, "We need to ask whether we are prepared to reduce the entire natural world to the status of an artifact."[14]

Members of the public will find themselves at different locations on the conservationist-preservationist continuum, with few positioning themselves at the far ends. Since both HBOs and NCs involve the manipulation of that which already exists (that is, nature), those who are on the preservation end of the continuum are expected to be more opposed to these technologies.

Industrial Approach to Nature and the Synthesized World

One way to determine where respondents lie on the conservationist-preservationist spectrum is to assess their views of an industrial approach to modifying nature. Should we only modify nature in a few experiments, or should we create self-perpetuating processes and risk creating a fully synthesized world?

With this in mind, I added factors to the HBO vignette to see if using industrial terms in describing the production of HBOs changed the level of support or opposition. Thus, 50 percent of the respondents read that scientists had created one HBO, and 50 percent read that they had created 10,000 HBOs, a figure meant to suggest the wholesale manufacture of HBOs and thereby to evoke concern about treating nature like a factory. But the results showed that this change had no effect on respondents' approval of HBOs.[15] Thus, either the theory that an industrial description drives opposition is not supported, or the respondents were not interpreting the description as I intended. Perhaps they thought that the 10,000 figure suggested that the safety of HBOs had been assured and therefore was not a cause for concern. We will leave this puzzle piece aside and see if it fits in with a broader picture later.

In another factor, for 50 percent of the respondents HBOs were described as "grown," and for 50 percent they were described as "manufactured"—again suggesting an industrial approach to nature. Seeing "grown" made respondents slightly more approving than seeing "manufactured"; 72 percent of those who saw "grown" somewhat agreed or strongly agreed that HBO research should continue, compared to 68 percent of those who saw "manufactured." This is a small difference, but a difference nonetheless, and it suggests that those who oppose an industrial approach to nature may be slightly more opposed to HBOs.

We can also look at the NC vignette. Half of the respondents read that scientists were creating a "new species," and half read that scientists were creating "no more than a dozen" animals. Those who read "new species" turned out to be more opposed to NCs[16]— not surprisingly, since "new species" suggests a more industrial approach to nature, with self-perpetuating consequences. But let us not lock this puzzle piece into place quite yet, because we cannot rule out another explanation: that respondents are less approving when they see "new species" because they fear making an error that could have a more profound impact on the environment.

Nature Has a Purpose and Plan

Those on the preservationist end of the continuum think that nature has intrinsic value and that we should, as much as possible, let it progress independent of us humans by its own logic, purpose, plan, or design. We should expect that such people would be opposed to the creation of HBOs and NCs, because clearly neither technology is part of nature's plan independent of humans. Thus, the survey asked whether "every species has its own purpose that it should be allowed to follow." Almost nobody strongly disagreed with this (.5 of 1 percent); only 3 percent somewhat disagreed, 15 percent took the middle position, while 43 percent somewhat agreed and 38 percent strongly agreed.[17]

We see a slightly greater approval of HBO research from those who think animals have their own purpose.[18] This small effect is in the opposite direction of what might be expected. But note that the proper interpretation of this question is difficult, because HBOs are not animals, and a good portion of the respondents think of HBOs as part of humanity, not nature, due to ephemeral connections.

On the other hand, views of species purpose strongly influence views on NCs, presumably because making the NC violates the purpose of the lab mouse or monkey. In this case, for those who most disagree that a species has its own purpose, 58 percent somewhat or strongly agree with continuing HBO research, and for those who most agree, only 33 percent somewhat or strongly agree with creating NCs.[19] This is a fairly large difference by survey standards.

Human Dominion over Nature

Another way of assessing a respondent's place on the conservationism-preservationism continuum is to directly ask about the conservationism end of the continuum—that is, about humans being in charge of nature. I measure this concept with two statements taken from previous studies: "Humans should rule over the rest of nature," and "Plants and animals exist primarily to be used by humans."[20] The responses to these statements are highly related, so I combined them into one measure of belief in "human dominion."[21]

The analysis shows that belief in human dominion is not associated with approval of HBOs.[22] Again, this is not surprising. While

HBOs are not "natural," they are perhaps considered as part of humanity, as we saw in the last chapter, whereas plants and animals are considered as part of "the rest of nature." In contrast, given that the survey questions ask about animals, it is not surprising that human dominion is strongly associated with approval of NC research.[23] Only 20 percent of those who agreed the least with human dominion somewhat or strongly agreed with creating NCs, while 63 percent of those who most agreed with human dominion somewhat or strongly agreed.

Part of the conservationist view is that humans should control nature, which requires understanding it. Those who think that we should understand all aspects of nature would then presumably be more supportive of HBO and NC research. I asked respondents to evaluate the statement "Some aspects of nature should remain mysterious and unpredictable." Twenty-one percent strongly agreed with this statement, and 29 percent somewhat agreed; 26 percent took the neutral position, and only 25 percent somewhat or strongly disagreed. The majority of the public evidently believes that some things should remain mysterious.

Believing that nature should remain mysterious and unpredictable is one of the strongest and most consistent predictors of views of HBOs and NCs.[24] Of those who most strongly think nature should remain a mystery, 27 percent somewhat or strongly agree with creating NCs, while of those who least think nature should be mysterious, 59 percent somewhat or strongly agree. For HBOs, of those who most strongly think nature should remain a mystery, 62 percent somewhat or strongly agree with HBO research, and of those who least think nature should remain a mystery, 86 percent somewhat or strongly agree with HBO research.

To sum up this section, the public's views of the relationship of humans to the natural world are strongly predictive of its views on HBOs and NCs. This is particularly true for views of NCs. I suspect this is because the public regards NCs as being a modification of something existing in nature (an animal), whereas HBOs are seen as unmodified parts of humans. In my survey, those who view nature as existing for responsible human use were more in favor of these technologies, and those who think nature should be relatively independent of us humans were less so.

The Cultural Authority of Religion and Science

Let me turn to other features of respondents that we can expect to be consequential for views of biotechnology in general and HBOs and NCs in particular. We can start by asking, What are HBOs and NCs *about?* The question can be answered in several ways: They are about human health. They are about "nature." They are about what we think about animals, what we think about humans, and, most importantly, what we think about the relationship between humans and animals.

There are two basic ways that Westerners have historically thought about similar subjects: through religion and through science. These two basic orientations have often been in conflict, although not in the way most people assume.

As a generalization of the entire scholarly field regarding the relationship between religion and science, we can say that, before the mid-nineteenth century, religion and science were rarely at odds.[25] Most of the scientists from this earlier era thought they were investigating the details of God's creation. Religion and science were often fused, as found in the common notion of the "two books of God," where "nature constituted one set of facts and the biblical Scriptures constituted another," and both books were seen to reach the same conclusion. This was not just the view of the religious but also the view of scientists, who tended to be religious themselves.[26]

From the late nineteenth century forward, however, religion and science diverged significantly, and while most fact claims about nature remain agreed upon by both (e.g., why flowers open in the morning), certain fact claims became contested by some religious groups (e.g., human origins, for conservative Protestants). But the "religion-science conflict" over facts about nature is no longer much of a conflict. What happened, particularly after the 1950s in the United States, is that Christianity more or less stopped making claims about the natural world that were different from scientists' claims.[27] Yes, there are conservative Protestants who reject evolution, but on any claim that matters to daily life, even they believe in science. The religion-science conflict turned from being about facts of nature to being about values, both concrete and more abstract. The topic of this book is a case in point: no religious person will claim that scientists have their *facts* wrong about HBOs and NCs, but many will say that they have their *values* wrong.

As an example of concrete values, there is a conflict between conservative Christians and scientists over technologies such as embryonic-stem-cell research. Part of the purpose of this book is to discover whether there is such a conflict over the concrete issues of HBOs and NCs. The developing understanding in the sociology of religion and science is that it is not only religion that has concrete and more abstract values, but also science. However, because science portrays itself as value-neutral, these values are often hard to identify.

What religious believers are really in conflict with are what they perceive to be the more abstract values articulated by scientists that underlie technologies like HBOs and NCs. The previous analyses in this book are all about these abstract values, including the human-animal distinction, the human-object distinction, the materialism-nonmaterialism distinction, and views of the proper human relationship to nature. These ideas, which scientists do have a position on, are not about facts but about values.

Competing Cultural Authority

I have not yet discussed the abstract value conflict of cultural authority. This conflict concerns which of the two realms—religion or science—has the cultural authority to be "in charge of" certain tasks in society. The modern description of this most abstract level of value conflict was best summarized by the late paleontologist Stephen J. Gould, who wrote that science and religion constitute "non-overlapping magisteria." In his summary, "Science tries to document the factual character of the natural world, and to develop theories that coordinate and explain these facts. Religion, on the other hand, operates in the equally important, but utterly different, realm of human purposes, meanings, and values—subjects that the factual domain of science might illuminate, but can never resolve."[28]

Similarly, Ian Barbour, who was probably the most influential modern scholar of the relationship between religion and science, calls this relationship "independence," in which religion and science "refer to differing domains of life or aspects of reality." Science asks "how things work and deals with objective facts," while "religion deals with values and ultimate meaning."[29] When either religion or science strays from its magisterium—that is, when the religious attempt to venture into fact claims, or when scientists attempt to venture into promoting values—it brings about an abstract value conflict.

The majority of American scientists agree with Gould's and Barbour's descriptions of the division of authority. A survey of American scientists, when asked about the relationship between religion and science, found that a majority endorsed the option "Independence: they refer to different aspects of reality" (50.9 percent) over the other three options: "Conflict: I consider myself to be on the side of religion" (0.4 percent of respondents), "Conflict: I consider myself to be on the side of science" (28.6 percent), and "Collaboration: each can be used to help support the other" (12.1 percent).[30]

But there has long been a very public faction of elite scientists who do not accept the division of authority in Gould's and Barbour's depictions and instead want *science* to be the source of values and ultimate meaning for society—to venture beyond science's magisterium. If we accept Gould's and Barbour's account, this would represent science taking over tasks usually performed by religion. To telegraph where I am going, people who agree with this expanded role of science are more supportive of the development of technology in general and have other distinct views of biotechnology.

One example of this culturally expansive role for science comes from the scientific atheists—most famously, Richard Dawkins, author of works such as *The God Delusion*[31]—who want to use science to show that religion is false. In other writings I have extensively described these scientific expansions into religious territory, which usually involve issues concerning the human body.[32] One of the most revealing statements from a scientist moving into the religious magisterium is that of Robert Edwards, the coinventor of in-vitro fertilization, who wrote that "many non-scientists see a more limited role for science, almost a fact-gathering exercise providing neither values, morals, nor standards. . . . My answer . . . is that moral laws must be based on what man knows about himself, and that this knowledge inevitably comes largely from science."[33] Similarly, the mathematician, public intellectual, and interpreter of science Jacob Bronowski wrote in 1962 that "I am, therefore, not in the least ashamed to be told by somebody else that my values, because they are grounded in my science, are relative, and his are given by God. My values, in my opinion, come from as objective and definitive a source as any god, namely the nature of the human being."[34]

Scientism

Another term that scholars have used to discuss an expansion of science beyond its magisterium is *scientism,* a term with varied meaning.[35] Gregory Peterson writes that scientism claims "that science can and should be the source of value and ethics" as well as "a source of meaning and purpose."[36] Peterson interprets scholars as concluding that scientism is the claim of salvation through science alone and "that the only kind of reliable knowledge is that provided by science, coupled with a conviction that all our personal and social problems are 'soluble' by enough science."[37] Another summary concludes that "scientism is roughly the view that science has no boundaries, i.e. that eventually it will answer all theoretical questions and provide solutions for all our practical problems."[38] Daniel N. Robinson writes that "scientism is, in its basic form, a dogmatic overconfidence in science and 'scientific' knowledge."[39]

Many scholars have described the interlocking facets of scientism. Philosopher Susan Haack, for example, lists six.[40] The details here are not necessary for my point, but we may take as exemplary Daniel Robinson, who lists four related facets: (1) "Only certifiably scientific knowledge counts as real knowledge. All else is mere opinion or nonsense"; and if 1 is true, then (2) "the methods and assumptions of the natural sciences, including epistemological and metaphysical doctrines, are appropriate for all science including, prominently, the social and human sciences"; (3) advocates hold "an exaggerated confidence in science . . . to produce knowledge and solve the problems facing humanity"; and (4) metaphysical claims are promoted, including the claim that "the world must really be like the methods of contemporary natural science assume it to be."[41]

From the view of scientism's critics, scientism is not science but rather an ideology, an attempt to "hijack science to support metaphysical commitments in which science has no particular interest." Indeed, "to resist this hostile takeover is to defend science."[42] The critique is that scientism uses the success and authority of science to promote a metaphysical and ethical agenda held by many elite scientists. The survey of scientists reported earlier suggests that most scientists are advocates not of scientism but of separate magisteria.

Scholars' lists of scientist advocates of scientism include Francis Crick, Richard Dawkins, Stephen Hawking, Carl Sagan, and E. O. Wilson.[43] A recurring theme is that religion and philosophy have failed, so it is time for science to take over. This was famously expressed by Wilson, who wrote that "scientists and humanists should consider together the possibility that the time has come for ethics to be removed temporarily from the hands of the philosophers and biologized."[44]

Steven Pinker is perhaps the best-known contemporary advocate of scientism, and reactions against his views are exemplary. According to Daniel Robinson, Pinker concludes that "science, as he describes it, provides the best foundation for belief, morality, and essentially all human endeavors," and that "the findings of science entail that the belief systems of all the world's traditional religions and cultures . . . are factually mistaken." Pinker largely derives ethics from science, writing that "the worldview that guides the moral and spiritual values of an educated person today is the worldview given us by science."[45]

If scientists set the values for society, what would those be? Explaining the particular ethics advocated by scientism would take an entire book. My empirical project cannot make assessments at this level of detail, so I have a much more basic explanation for the connection between belief in scientism and the ethics of HBOs and NCs. In debates over issues portrayed as moral or ethical, such as HBOs and NCs, people who believe that science is the solution to all problems and that science itself provides ample basis for human ethics will assume that whatever scientists are doing is acceptable.

I have produced one set of analyses focused on religion and one on the cultural authority of science, and have sought to determine whether religious people are more opposed to HBOs and NCs than are the nonreligious. If this were the case, it would be religious opposition similar to the disproportionate religious opposition to scientific efforts to modify the genetics of humans.[46] I will also see if those who think they are in moral conflict with science, and those who oppose scientism, are less supportive of creating HBOs and NCs.

Which Religious Believers Are Supportive of and Opposed to HBOs and NCs?

Since religion has a history of moral opposition to scientific projects that involve the human, and biotechnology in particular, we can predict that the religious will be more opposed to HBOs and NCs than the non-religious. I want to examine this question in much more detail, asking *which religious traditions* and *which particular beliefs* are correlated with more or less support. It will also be important to closely examine religion because it is one of the few possible sources of organized opposition to these technologies. While I can imagine animal-rights organizations and more radical environmentalists being opposed to HBO and NC technology, there is also a strong history of religious groups being opposed to scientists' efforts on issues that involve the definition of human beings, such as human embryos.

If you have been following the news media in recent years, you may well have become convinced that religious believers in the United States are largely theologically conservative and right-wing politically. In actuality, and in contrast to many other countries, Christians in the United States exist across the whole theological and political spectrum. Some are politically conservative and involved with the anti-abortion and anti-gay rights movements. Some are politically liberal and involved with the prochoice and pro-gay rights movements. Religious liberals have also long had key roles in the antiwar, civil-rights, and other similar movements.[47]

The dominant religion in the United States is Christianity, but there is wide variation in how the different traditions within

Christianity view God, science, nature, the human, and animals. For example, liberal Protestants (e.g., Episcopalians and Congregationalists) are often more trusting of science and scientists than are the nonreligious. Among US Christians, conservative Protestants are typically further to the right than Roman Catholics, though the views of Catholics do not necessarily map neatly onto the Protestants' categories.[48] Understanding this variation within Christianity is important for any sociological study of biotechnology.

In a study such as this, it is important to distinguish between people who are actually committed to a religion and those who merely retain some sort of residual religious identity from their childhood. The categories of religiously committed respondents as identified in the survey are: Catholic (13.2 percent of respondents), conservative Protestant (21.9 percent), liberal Protestant (5.2 percent), identity-rejecting Protestants (7.0 percent), and members of other religions (7.2 percent). I also identify low-commitment Christians (23.6 percent),[49] and the nonreligious (21.7 percent). (For details of how this categorization was done, see the Methodological Appendix.[50]) I will not discuss the "other religions" category from here on because it is uninterpretable, comprising a number of distinct traditions (e.g., Muslims, Hindus, Jews) that each amount to 2 percent or less of the US population, too small a percentage for statistical analysis.

The three largest theological and denominational traditions in the United States are Catholicism, conservative Protestantism, and liberal Protestantism. Between the late nineteenth century and the 1930s, Protestants became less split by specific tradition (e.g., Baptists, Presbyterians, Lutherans) and more split over orientation toward knowledge. The use of Enlightenment rationality in biblical interpretation produced the debate over "higher criticism" of the Bible, also known as the fundamentalist-modernist controversy. This latter dispute involved, for example, biblical modernists who, relying on modern discourse analysis, suggested that the book of

Isaiah was actually written by two people, while fundamentalists asserted that it was written by a single man named Isaiah. These two groups also had different orientations to science, with the modernists deferring to scientists' claims and the fundamentalists retaining traditional interpretations of biblical fact claims about the world.[51]

In the 1940s the fundamentalists had a schism where a group called evangelicals broke off, with the fundamentalists being more literalist, more skeptical of science, and more isolationist than the evangelicals.[52] Both groups were opposed to teaching evolution. For my purposes here, I will not further distinguish between the two but will call them collectively "conservative Protestants," as there are not enough fundamentalists to separately analyze using a survey.

The modernists have come to be called "mainline" or "liberal" Protestants. This tradition arguably exists to use Enlightenment reason in its interpretation of the Bible and therefore is appreciative of science, which is also built on Enlightenment reason. While Protestantism is represented by Lutherans, Presbyterians, Baptists, Methodists, and so on, the individual traditions of these denominations do not tell us much, as each is split to varying degrees between conservatives and liberals.[53]

Catholicism is one of those religious traditions (like Judaism) that holds that if theology makes one claim and science another, then either one of them could be wrong. Due to Catholicism's rigid institutional structure, it can take a long time for a scientific claim to be accepted into the official teaching of the Church. For example, it took over 100 years for the Catholic Church to state clearly that it agreed with Darwin's description of human evolution.[54]

Because of changes in American religion and society since the mid-twentieth century, these different types of Christians all now agree with scientific methodology and defer to scientists' claims for almost all fact claims about nature.[55] Though conservative Protestants will still say that they do not agree with scientists on claims about human origins, including evolution, they are committed to science on any issue that actually matters for an ordinary citizen's life. I would be hard-pressed to find any members of the public who would claim to disagree with scientists' description of how an HBO grows or whether the human tissue implanted in a mouse brain is active. Despite a few factual disputes, the primary issues that arise between religion and science are value differences.

On concrete issues like HBOs and NCs, liberal Protestants can be expected to have the same values as scientists. Depending on the issue at hand, Catholics are sometimes in value conflict with science, but it is conservative Protestants who are most consistently in value conflict with science.[56] Thus, one would expect Catholics and especially conservative Protestants to express less approval of HBOs and NCs than liberal Protestants.

"Identity-rejecting Protestants" are a newly emerging phenomenon in survey research. These are people who scholars would categorize as Protestants but who do not identify with the term *Protestant* or with any of the other identity labels that American Protestants use (e.g., *mainline, born-again,* or *evangelical*). They are, however, religiously committed; I only include people in this category if they also have other markers of engagement and conservative Protestantism. In a different study I found that this group was distinct from other groups, and I consider it to be a populist, anti-institutional version of conservative Protestantism.[57] In the Methodological Appendix I further investigate this group, showing that they are younger, less white, less educated, and much more politically liberal than conservative Protestants—despite being in agreement with conservative Protestants on theological beliefs such as biblical literalism.

The nonreligious are increasingly seen as a distinct "religious" group and not simply a residual category representing the absence of religion.[58] This group includes "atheists," "agnostics," and "nothing in particular." While these categories are to some extent distinct, I have combined them for the sake of parsimony and because this is not a book about varieties of nonbelief. In my analyses, the nonreligious are always the reference group; thus, whenever I write, for example, "Conservative Protestants are more opposed to X," this actually means that "Conservative Protestants are more opposed to X than are the nonreligious."

Impact of Religious Tradition on Views of NCs and HBOs

I conduct distinct analyses to describe and explain the views of religious groups. Description simply shows if members of particular religious group are supportive or opposed to the technologies. With description, why a religious group is in support or opposition is not considered. But, to truly understand the perspective on these technologies, the *why* is important. This is explanation. I will therefore move back and forth between a description of the views of a group and a (partial) explanation.

Description, in this case, answers the question of, if there were 100 conservative Protestants in a room, how many of them would be opposed to HBO research? *Explanation* would require answering questions such as, Is it their religion that leads to this opposition, or some other social feature that conservative Protestants tend to share? Though many of the commonsense ideas about which groups are opposed to science are descriptively true, it is typically not actually membership in a given group that leads to opposition but rather some other trait that members tend to share. For example, saying that conservative Protestants are less likely than other groups to believe scientists' claims about climate change is an accurate description. But an explanatory analysis would reveal that what typically leads conservative Protestants to reject climate change is not their religious beliefs but their immersion in politically conservative discourse.[59]

We may begin with description: The estimated proportion of the nonreligious who would select either "Somewhat agree" or "Strongly agree" to the statement "research into HBOs should continue" is 78 percent.[60] This group is very supportive. The analysis also predicts that 70 percent of Catholics would select either "Somewhat agree" or "Strongly agree" to HBO research, reflecting a statistically significant but not substantively large difference between Catholics and the nonreligious. Liberal Protestants express, statistically, the same level of support as do the nonreligious—a common finding in research on biotechnology. For conservative Protestants, the somewhat or strongly agreeing rate was 63.2 percent, and for identity-rejecting Protestants it was 53.6 percent—figures substantially different from those of the nonreligious.

But would this opposition be due to what they hear in church? Religious people have nonreligious beliefs and experiences that influence their attitudes as well. As many evangelical Protestant religious leaders have bemoaned in the age of Donald Trump, the pastor may only have the ear of the faithful for 30 minutes per week, but Fox News has their ear for 12 hours.[61] This suggests that other experiences of members of particular religious groups may be more important than religion per se.

For explanation, we return to adjusting the statistics for all the characteristics of the respondent and views of nature described earlier in this chapter, as well as the attitudes toward science questions I will describe below.[62] In this analysis, the identity-rejecting Protestants are more opposed.[63] When we incorporate all the features of religious respondents that could actually be producing opposition (e.g., different education levels, being more Republican, or having a different view of science than others), the analysis suggests that there is indeed something about identity-rejecting Protestantism that leads to disproportionate opposition to HBOs.

I conducted a parallel analysis of approval of NC research. Regarding description, all the religious groups express the same level of approval as the nonreligious except the liberal Protestants, who are more supportive of NC research than are the nonreligious.[64] Regarding explanation, our results show that attitudes of the religious are not driven by their religious beliefs per se.

In Chapter 2 I cited social theorists who have concluded that Christianity teaches foundational distinctions in general and the human-animal distinction in particular, so it is surprising that religion seems to have so little influence on views of NCs. Keep in mind that these are all comparisons, so the lack of difference with the nonreligious may be because the nonreligious have some nonreligious reasons why they are opposed to NCs—for example, a strong belief in animal rights, or serious concern about environmental accidents. To figure out what is really happening will require further research.

Theological Beliefs Crosscutting Christian Traditions

While readers uninterested in American religion may think that the above categories are fine-grained enough, I also want to investigate whether a number of classic theological debates influence views of HBOs and NCs. As we have just seen, religious identities may turn out not to be important, and what may instead be important are theological beliefs that cut across traditions, beliefs that are directly relevant to abstract values such as views of humanity, animals, and nature. An explanatory analysis will seek to reveal whether these beliefs lead to support or opposition to HBOs and NCs.

The first theological idea that is relevant to this debate and that cuts across Christian traditions is the extent to which the respondent primarily believes that a human is defined by being made in the image of God. A traditional implication of this idea is that something invented or created by humans is not human. Since Adam and Eve were created in God's image, entities that are not the biological descendants of that original couple (e.g., NCs) are not. More relevant for NCs, animals have traditionally not been thought of as being in the image, although in Catholicism this remains a bit ambiguous. I have a survey measure of the extent to which respondents believe in this Christian theological anthropology.[65]

While questions about the definition of a human were asked of all respondents, some respondents were also asked three theology questions. These were so specific that only those exposed to Christian teaching would have a position on them; therefore, they were only asked of self-identified Christians, to whom the underlying concepts would be recognizable even if they were unlikely to use the words in everyday conversation.

The analysis shows that the more respondents believe that humans are made in the image of God (regardless of which religious group they belong to), the more likely they are to be opposed to HBOs as well as NCs.[66] Comparing those least likely to believe in this anthropology and those most likely, the portion of each group strongly agreeing or somewhat agreeing with HBO research was 70 percent and 62 percent, respectively. Differential belief in the theological anthropology (that humans are defined by being in the image of God) among Christians also resulted in a 12-percent difference in selecting "Somewhat agree" or "Strongly agree" with NC research. These differences for HBOs and NCs are not large.

The second theological idea is "dominion beliefs" that form the basis for the way many Christians approach the natural world. These ideas were most famously identified by Lynn White, who claimed in 1967 that monotheism and the Genesis creation account in Judaism and Christianity divested nature of divinity and told humans to subdue the earth. This anthropocentric view of the world legitimated the environmental exploitation of the planet.[67] Critics of White's thesis immediately pointed out that the book of Genesis also authorizes a different interpretation, since God tells humans there to be the stewards of creation. Research on whether dominion beliefs lead to more or less environmental concern remains inconclusive.[68]

Therefore, my survey attempted to measure the specifically Christian version of this dominion belief by means of this statement: "The best understanding of the first chapter in the book of Genesis in the Christian Bible is that: (a) God told humans to use nature for human goals; (b) God told humans to guard and supervise nature." My initial expectation was that those who think God told us to use nature for our goals should be more supportive of HBO and NC research because we are indeed using the natural world for our benefit. However, the analysis shows that dominion beliefs are not related to respondents' views of HBOs. Again, this may be because HBOs are not considered to be part of nature but rather part of us humans. However, dominion beliefs do predict views of NCs. Among those who thought nature is intended to be used for human goals, 50 percent somewhat approve or strongly approve of NCs. Among those who believed we are obligated to guard and cultivate nature, 38 percent would somewhat or strongly approve of NCs. Unlike HBOs, NCs are considered part of nature, in that they involve animals. For someone who believes we are free to use nature for our goals, creating NCs for our own medical research needs does not present an issue. If someone believes we are obligated to guard nature, on the other hand, modifying nature is not guarding.

The third theological concept is whether humans are primarily creatures created by God or are cocreators with God.[69] People who believe in the former are generally theological conservatives, and those who believe in the latter generally liberals. If we are creatures and only part of God's creation, then it is not our role to modify the existing design of nature by creating HBOs and NCs. If we are instead cocreators, then God gave us the brains and wisdom to complete creation, which could include modifying humanity and animals. Most theologians fall in-between these end points, so the laity could be expected to as well.

To measure this concept, the survey stated: "There are many ways to compare the responsibilities of humans and the responsibilities of God for the natural world. One continuum is below. Please place your views upon that continuum." On one end was the statement "Humans are to *oversee* what God has already created"; on the other was the statement "Humans are to *aid* God in God's ongoing creation."[70] The responses of Christians turned out to be very evenly distributed.

In the HBO analysis, seeing the role of humans not as cocreators but as overseeing God's creation was associated with opposition to HBOs.[71] But the difference in approval of HBO research between those on the two ends of the continuum was very small, a mere 1.5 percent. As for NCs, there was no difference in approval between the people on the two ends of the spectrum. Thus, this theological idea has essentially no impact, at least when the statistical analysis simultaneously considers the other theological ideas.

A fourth cross-tradition theological concept in discussions of religion and science is eschatology—the trajectory of history and humanity. In Christianity, eschatology includes the second coming of Christ and the nirvana-like Kingdom of God at the end of time. A central question is how that paradise will come into being. One perspective is post-millennialism, in which we humans create the Kingdom of God on Earth through social improvement, after which the second coming of Christ occurs. This was the motivation for much liberal social action by Christians over the past few hundred years and was most evident in the American Social Gospel movement of the early twentieth century. The opposite end of the spectrum is pre-millennialism, associated with theological conservatism, in which the Kingdom is expected to come into being via divine action independent of human efforts. In pre-millennialism the Kingdom is totally supernatural in origin.[72]

As in the previous question, those who believe that Christians can bring on the Kingdom through improving the world should be more likely to think that modifying Creation with HBOs or NCs is allowable or even obligatory. My survey statement read, "It is often said that Christians yearn for the Kingdom of God on Earth. Please indicate on the continuum below whether you think the Kingdom of God will" [at one end] "come into being largely independent of human effort" or [at the other] "be built through the efforts of Christians."[73]

There was a very large difference in approval of HBOs between those who see humans as bringing on the Kingdom of God and those who see this occurring independently of human action. Of those who think we should wait for God to bring on the Kingdom, 56 percent of respondents somewhat or strongly agreed that HBO research should continue. Of those who think we humans should bring on the Kingdom by improving the world, 74 percent were somewhat or strongly approving of HBO research.

As for NCs, those who believed the Kingdom would occur independently of human action were strongly inclined to oppose NCs. Of those who see the coming of the Kingdom as independent of human effort, 32 percent somewhat or strongly agreed that they support the creation of NCs, whereas of those who think the Kingdom requires human effort, 50 percent somewhat or strongly agreed. Again, belief that humans have a religious obligation to improve the world results in more approval of research.

This turns out to be the most influential theological idea in my analysis. (Why that is the case would require a separate study.) This distinction, a factor in the divorce between the fundamentalists and the modernists, has long been one of the more consequential theological divides, at least within Protestantism.

This may well be one of the actual divides within Christianity over science. That is, the ability to generate fact claims about the world is not a very important divide. The most important divide is the necessity of using our human ingenuity and effort—largely through science and technology—to improve Creation and bring on the Kingdom.

To sum up: If we examine the entire public, religious and nonreligious, we can see the effect of being nonreligious in comparison to being committed to any of a range of religious traditions. As a description, we see that the two types of conservative Protestants, as well as Catholics, are more opposed to HBO research. No group is more opposed to NCs compared to the nonreligious.

In the explanatory analysis, I examined the impact of religious tradition by removing the influence of a number of other features of the respondent. Beyond demography, this includes views of nature, views of the distinction between humans and animals, and views of scientific authority. Except among one type of conservative Protestant, these views of scientific authority, nature, and the human-animal distinction account for all of the variation in approval of HBO and NC research. This means that, to the extent that religious groups are differentially opposed to HBOs, it is because their members are disproportionately from certain demographic groups and have particular views of the human relationship with nature and the cultural authority of science.

I have identified some theological ideas within the Christian community that seem to determine approval of HBO and NC research. In the late 1980s, the sociologist of religion Robert Wuthnow made the then-novel point that the differences between denominations (e.g., the United Methodist Church vs. the Southern Baptist Church) were not as stark as the differences between liberals and conservatives within each denomination.[74] Something similar seems to be the case here, probably constrained to analyses of religion and science or religion and nature, which is that each tradition is not divided by liberals and conservatives so much as by theological concepts such as eschatology. Future survey researchers should consider asking survey questions that measuring these concepts so they can see the actual influence of religion on beliefs about biotechnology.

The Expanded Cultural Authority of Science

The flip side of faith in religion is faith in science. As mentioned earlier, religion and science are not in meaningful competition over explaining, for example, how an organoid works; almost everyone would agree that understanding the physical world is the domain of science. Rather, the larger issue is whether the public today looks to science as a source of meaning and societal direction—that is, for functions that in previous decades would have been fulfilled by religion. In the terms I use above, one's endorsement of expanding science's cultural authority beyond its magisterium may be associated with approval of NCs and HBOs.

The Republican War on Science

I am focused on the possibility that those who believe in the cultural authority of science are those who are most supportive of HBO and NC research. But in my study I want to account for traditional explanations for why the public does or does not support a scientific technology. A common and much more general explanation for variation in support of science is that political parties have influence on people's views. One of the repeated findings in the sociology of science over the past 20 years is that political conservatism, and the Republican Party in particular, has been increasingly hostile to science.[75] Recent research has clarified that the hostility is not necessarily to science but to scientists.[76]

My study, like many other studies of the public views of science, shows that Democrats are somewhat more supportive of a specific scientific technology than are Republicans.[77] For example, only 62 percent of Republicans somewhat agreed or strongly agreed that HBO research should continue, while 75 percent of Democrats somewhat or strongly agreed. Similarly, 37 percent of Republicans somewhat or strongly agreed that they support creating NCs, compared to 46 percent of Democrats. However, when I turn to explanation—to whether the effects are consequences not of political ideology itself but rather of characteristics that Republicans tend to share, such as their age and race—there is no difference in level of approval of HBOs and NCs.[78] What might appear to be the influence of political party membership actually turns out to be the influence of other attributes such as religion and scientism, which I will address below.

Levels of Scientific Knowledge

A traditional assumption, especially among scientists, is that lack of support for scientific research is due to the public's not understanding the science itself, and that if the public just understood the science, especially on issues like HBOs and NCs, it would be supportive.[79] This presumes that people's ignorance makes them fearful of science that they do not understand. But this perspective, called the *knowledge deficit model,* is often found to actually be a fairly small contributor to opposition to scientific developments.[80] The claim that it is the primary or most important factor is now considered one of the great myths of science communication, one which has been extremely difficult to counteract.[81]

To account for this explanation, I gave respondents a self-assessment statement, providing a "Strongly agree"-to-"Strongly disagree" scale for their responses: "I am informed about science and technology." Only 48 percent of those claiming the least knowledge of science somewhat or strongly agreed with continuing HBO research, compared to 80 percent of those claiming the most knowledge. Similarly, when asked about supporting the creation of NCs, the corresponding numbers were 15 percent and 58 percent. These are very large effects, and they seem to suggest that approval of these technologies is importantly driven by knowledge of science.

But is it respondents' scientific knowledge that led to their views of these technologies, or some other predilection that scientifically knowledgeable people tend to also share? The answer is important because, if the difference depends literally on knowledge, then proponents of these technologies should presumably engage in a public campaign to explain how HBOs are formed and kept alive, and how human tissue is actually implanted in another animal.

As I have done in previous analyses, I then adjusted the figures, taking account of other characteristics that the more knowledgeable respondents share. Many of these—most notably educational level—are probably associated with knowledge of science, and may be the principal reason for supporting research. This analysis reveals that approval of HBO research does not depend on actual knowledge of science.[82]

For NCs, the extremely large descriptive effect of knowledge is attenuated by other factors, but it is nevertheless evident that knowledge of science *does* have a modest impact on views, with a nearly 14-percent difference between the low- and high-knowledge respondents in their selecting "Somewhat agree" or "Strongly agree" with the creation of NCs.[83] So, while it is correct that people with more scientific knowledge are much more supportive of technology, it is generally not the knowledge itself that leads to this effect, but rather something else that scientifically knowledgeable people tend to possess.

Shared Morals and Trust in Science

I turn to an attitude toward science that has rarely been investigated by scholars. In recent years it has become clearer that science is in moral conflict not only with religion but probably with the public in general. Therefore, approval of biotechnologies like HBOs and NCs may be determined by whether respondents perceive that scientists are not engaged in moral behavior. To determine whether respondents think they are in moral conflict with scientists, I asked them to evaluate the proposition "Scientists and I have similar morals."

Those who believe they do not inhabit the same moral community as scientists are unlikely to trust them. If scientists do not share your values, the reasoning goes, why should you trust them to do what is right? To evaluate this perception, I asked, How much confidence, if any, do you have in scientists to act in the best interests of the public? As expected, the morals and trust issues turned out to be so correlated that I combined them into a single measure.[84]

Belief in shared morals and trust are clearly key for views of HBOs and NCs. Of those at the bottom of the trust science/shared morals scale, only 36 percent somewhat or strongly agree that HBO research should continue, whereas 90 percent of those at the top of the scale have that level of approval. There is a near-identical 58-percent gap in agreeing with the creation of NCs. Morals and trust represent not only a description but an explanation; even when controlling for all other possible explanations, there remains a gap of approximately 55 percentage points in support of HBO and NC research between those at the top and bottom of the trust science/shared morals scale. This is clearly a very important driver of attitudes toward HBOs and NCs.

Scientism: Cultural Authority beyond Science's Magisterium

Scholars are coming to recognize that, for many Americans, science is taking on the social functions that were previously performed by religion, and having faith in science as a source of meaning in society may be what drives many people's view of scientific technology. This is known as *scientism*. One way I sought to measure adherence to scientism was through a statement about science's role in society: "Science should set society's goals." Another aspect of scientism is the belief that, if a question cannot be answered with science, it is not really an important question; thus, respondents were asked their opinions on another statement: "The most important questions for society can be answered with science." Another facet of scientism is the belief that science will solve our problems and be the engine of future human happiness. Thus, I presented another statement, one which had been used in surveys for decades: "Because of science and technology, there will be more opportunities for the next generation."

As expected, the responses to these three questions were so correlated that I combined them into an additive "scientism" measure.[85] If we look at the descriptive analysis (analogous to talking to 100 people in a room), there is about a 50-percent difference between those who most agree and most disagree with scientism on endorsing research with HBOs and NCs. Thus, in the metaphorical room of advocates of scientism, 92 of those you talk with would be in favor of HBO research, whereas in the anti-scientism room only 42 would have that view. The gulf remains similar (a gap of 42 percent for HBO and 40 percent for NC) when controlling for other possible explanations.

When scholars look at attitudes toward science to explain attitudes toward biotechnology, they rarely consider scientism. But these findings suggest that they should, as it is clearly a powerful explanation for attitudes toward biotechnology.

Which General Feature of Respondents Is Most Important?

As a final analysis, we can ask which characteristic of respondents are more important for their views of HBOs and NCs. We can calculate a measure that is basically the importance of a characteristic in determining the opinion about HBOs and NCs that in theory could go from 0 to 1, with 1 being the most important.[86] This is not terribly precise, so only larger differences should be considered. The effect of the demographics, party identification, and scientific knowledge are all extremely small, generally below .07. For example, the influence of age as a predictor of views of NCs is a mere .01. Much more surprising is that religion is similarly unimportant, falling in a similar range. The most important religious contributor to opinion is the response to HBOs of identity-rejecting Protestants, which comes in at .11.

By contrast, respondents' views of nature turn out to be very important. For NCs, all three measures of respondents' views of nature—the idea that some things should remain mysterious (.20), that humans should exercise dominion (.26), and that animals have their own purpose (.13)—were found to be highly influential. For HBOs, by contrast, the only strong influence was the idea that some things should remain mysterious (.20). Shared morals and trust also were shown to be fairly important, averaging .13 between the HBO and NC analyses. Agreement with scientism turned out to be similar in importance to attitudes toward nature, averaging .20 between the HBO and NC analyses.

Conclusion

In Chapter 4 I documented the impact of the public's definition of a human on its view of HBOs and NCs. It is definitions of a human that are responsible for the human-animal distinction and the public's belief in ephemeral connections to body parts. The impact of these concepts may be very specific to the HBOs and NCs, although we may find something similar for other technologies that involve the merging of humans and animals or for the use of disembodied parts of humans.

This chapter has considered HBOs and NCs as instances of biotechnology, focusing upon the more general characteristics of the public that can be expected to lead to public support or opposition.

In investigating the public's views of these technologies, it turns out that the respondents' demographics—including, surprisingly, their religion—are not very important. Political identification and amount of scientific knowledge are also apparently not very important. Instead, what we find is most important is the perception of nature, as well as the extent to which science is viewed as a meaning-making entity with cultural authority, akin to a religion. These are the general social attitudes that are driving the public's views of HBOs and NCs.

6

What Is to Be Done?

Scientists are in their labs creating improved HBOs and NCs. Bioethicists have been publishing many articles on the ethics of both technologies, and scientific associations such as the National Academies of Science have called for public input to deliberations on what to do about these technologies. However, any sort of public deliberative process is unlikely to begin any time soon.[1] Therefore, as a stand-in for the public we need empirical studies of the public's views of these technologies.

Many studies would just assess the extent to which the public agrees or disagrees with a technology. Having done such an assessment as a first step, I would say that the public is quite supportive of HBO research, at least as I have described it, but has much more mixed feelings about NCs. Unfortunately, knowing the level of agreement or disagreement alone has very limited usefulness in a public debate; what we need to know is *why* Americans, and particular subsets of the American public, are or are not opposed. If scientists knew why people are opposed, they could possibly alter the technology to make it more acceptable. Policymakers would also know where to draw moral limits based not on yes/no views of the technology but on the actual moral concerns of the public. In this book I have attempted to assess the "why."

When using a social survey to examine the public, it is often necessary to infer the "why" from the "who." For example, if women were found to be more opposed to a technology than men, we would infer that something about the experience of being female in contemporary society leads to particular views of technology. The "why" would then attempt to identify that experience.

Disembodied Brains. John H. Evans, Oxford University Press. © Oxford University Press 2024.
DOI: 10.1093/oso/9780197750704.003.0006

I have examined many of the characteristics that scholars might suspect could lead to distinct views of biotechnology in general—gender, race, education level, level of scientific knowledge, party affiliation, and religion—but none of these actually turn out to be very predictive of people's views. While some distinct theological ideas within Christianity tend to lead to particular views of these technologies, theological ideas are primarily relevant to the subpopulation of Americans who are active Christians.

Studies of the public's views about biotechnology do not usually examine the impact of beliefs about the natural world. I have described these views as a general dichotomy between the conservationist and preservationist views of nature. The conservationist wants to responsibly use nature for human ends, whereas the preservationist believes that nature has its own intrinsic value independent of human interests. Those who lean toward the preservationist end of the spectrum believe that existing nature has a purpose and value and are therefore reluctant to modify it with either HBOs or NCs. These beliefs have a very large effect on attitudes toward HBOs and NCs.

Most survey analyses of the public's attitudes toward science presume that science is concerned with discovering facts about the natural world. However, science also uses and implicitly teaches a moral perspective that the public may not agree with. Attitudes toward HBOs and NCs, it turns out, are partly driven by the belief that scientists do not inhabit one's own moral community.

Beliefs about the cultural authority of scientists have an enormous impact on perception of technology. Respondents who want science to take on the meaning-making functions typically embraced by religions are the most supportive of these technologies. The logic here is that if we should let scientists set the goals of society, and since scientists are creating HBOs and NCs, these must be included among society's goals.

I also focused on the public's distinct definitions of a human, which tend to support a foundational distinction between humans and animals. The public's definition of a human also leads to belief in ephemeral connections to disembodied human tissue. Such a belief tends to result in HBOs being seen as violating the foundational distinction between humans and objects. If an HBO still contains the "essence" of a particular human, then any experiments on HBOs will be seen to treat that (partial) human like an object.

These views are in contrast with the beliefs of bioethicists and scientists. The primary definition of the human implicitly used in the academic bioethical literature leads to the moral concern that an HBO or NC could obtain human-level consciousness, or at least enough consciousness that we would be required to treat it differently from the product of an ordinary lab experiment. This concern with consciousness easily flows from the underlying beliefs of bioethicists and scientists. I actually find little evidence that level of consciousness is a critical issue for the public, except at the extremes.

In contrast, what strongly determines the public's views regarding NCs turns out to be belief in a foundational distinction between humans and animals. Another strong determinant turns out to be a widespread rejection of the idea of humans as ultimately consisting only of chemicals. Whereas scientists would generally say that we are *only* material body and material mind, the public believes that humans are also spirit or soul. Whereas for the scientist the organoid is just a material set of cells, for many in the public the organoid has an ephemeral connection to the human who donated the cells. These differences in beliefs are associated with concerns about HBOs.

Cultural Impacts of HBO and NC Research

The public-policy bioethical debate is largely about the treatment of individuals. For HBOs, the question is whether we are mistreating an individual HBO, if it is aware, by keeping it in a dish. Is the individual NC mouse with increased mental capacities likewise being abused if we are not treating it more like we would treat a monkey?

A different tradition in the public-policy bioethical debate, which fell out of prominence many decades ago, is concerned less about individuals and more about cultural change.[2] In this tradition, the primary cultural issue that has concerned scholars is that technology could change the culture's definition of a human, and that that change would have pernicious effects on people as a whole. This is what the term "*de*-humanization" means: using the wrong cultural definition of the human when deciding how to treat people.[3] For example, Jason Scott Robert and Françoise Baylis are concerned about the "moral confusion" that could result when the creation of NCs slightly redefines humanness and in turn leads humans to treat each other slightly more like we treat animals.[4]

Another concern is that humans will be thought of as slightly more like inanimate objects, which violates the human-object foundational distinction. This concern is reflected in the extensive writing about "objectification" of people.[5] An object has no interests and can be treated purely as a means toward a person's ends, just as we would treat a chair. Obviously, the history of humanity has often departed from the humane ideal, and many humans have been treated like animals and objects. But an ideal it remains, exemplified through concepts like human rights.[6]

Of course, these concerns are not new. A primary concern of Darwin's nineteenth-century critics was not that Darwin had his facts wrong about humans evolving from animals, but rather that learning there was no foundational distinction between humans and animals would have a negative impact on human morality.[7] This is indicated by the famous quip by his contemporary Lady Ashley, who said about Darwin's theory, "Let's hope that it is not true; but if it is true, let's hope that it doesn't become more widely known."[8] Critics feared that Darwin was teaching us that we were subject to survival of the fittest, which would lead to immoral acts against others.

For example, William Jennings Bryan, who famously defended the creationist view in the Scopes "monkey trial" in 1925, was concerned with defending biblical literalism, but also concerned about, in the words of historian Ronald Numbers, "the paralyzing influence of Darwinism on the conscience. By substituting the law of the jungle for the teachings of Christ, it threatened the principles he valued most: democracy and Christianity." Darwin had taught us that we humans are more like animals, and Bryan thought that learning about Darwin's view of humans had led to the German decision to declare war in 1914.[9] To take another example, in the 1940s British author C. S. Lewis wrote that "once the old Christian idea of a total difference in kind between man and beast has been abandoned, then no argument for experiments on animals can be found which is not also an argument for experiments on inferior men."[10]

There is empirical evidence that such dehumanization followed by bad treatment does occur, at least at the extremes. Particular groups of humans have been redefined out of humanity by discursively equating them with animals and objects. For example, Sam Keen writes that "as a rule, human beings do not kill other human beings. Before we enter into warfare or genocide, we first dehumanize those we mean to 'eliminate.' Before the Japanese performed medical experiments on human guinea pigs in World War II, they named them maruta—logs of wood." Jews had to be redefined as "vermin" before the Holocaust, and enemy soldiers are typically redefined as animals.[11] More generally, Kristen Renwick Monroe has concluded, in a summary of a study of genocides, that "this initial psychological distancing from former friends and fellow citizens facilitates an eventual process of dehumanization of 'the other'—a cognitive perception that is the match that lights the tinder of genocidal violence. Only when friends and fellow citizens are dehumanized does the unimaginable become possible."[12]

Psychologists have examined how people are treated in the context of comparing humans to animals. For example, the psychology literature asks "how viewing others as nonhuman allowed us to morally 'disengage' from them—justifying treating them as animals [and] undermining the legitimacy of their views and needs." Critically, the literature focuses on the fact that "dehumanization is not just restricted to extreme or overt prejudice but can occur subtly and even without conscious awareness."[13]

Such nonconscious dehumanization is what would be the most plausible for NCs and HBOs. If, by learning about NCs, the public is taught that animals and humans are similar, then the public will think of humans as ever so slightly more like animals and treat them as such.

Some bioethicists who acknowledge this mechanism of cultural change hope to avoid any such potential mistreatment of humans by first making it unacceptable to mistreat animals. Henry Greely, who is generally supportive of creating NCs, writes that the most coherent argument he has seen against "humanization" of animals is the "moral confusion" theory of Robert and Baylis. He identifies the hypothesized impact of the confusion as "the fear" that it would lead people "to treat humans worse."[14] But Greely's hope is that familiarity with NCs may instead lead people "to treat nonhuman animals better."[15]

Indeed, academics who believe in animal rights argue that treating animals more like humans is a reason to remove the foundational distinction between humans and animals. Reacting to the "moral confusion" argument, Robert Streiffer writes that, "even supposing that chimeric research did cause society to revisit its views about the comparative moral worth of humans and nonhumans, it still remains that this would provide little reason not to create chimeras. Indeed, it seems more plausible that this would provide, on balance, a reason to create [NCs]." Streiffer goes on to say that humans torture and eat countless animals, and concludes that "revisiting the moral views that are invoked to support this situation is probably essential to improving this social institution."[16]

While Greely and Streiffer are correct that it is possible that the merging of animal and human in the public mind could improve our treatment of animals as they become defined as more human, this seems socially unlikely given how we treat animals at present.

There is an extensive literature going back to at least the nineteenth century on the impact of weakening the human-animal distinction. However, organoids are so new that little has been written about how mistreating disembodied parts could lead us to be less concerned about mistreating the original human. That said, in a discussion about scientists showing respect for human tissue that could be put in an animal, Greely states that "it is not clear whether these prohibitions [on how to treat human remains] stem from respect for the individual whose body parts or tissues are involved or from a fear that such uses hold humanity itself in disrespect— and may, in time, lead to even more noxious disrespect for living human persons."[17]

The concern is that HBOs will teach us that we are nothing but objects, to be treated accordingly. This would occur when parts considered to be "human" (as in the label "human brain organoid") are put in a dish and used like objects. Given that these objects retain an ephemeral connection to an actual human, they could be seen, from the perspective of people who believe in something like a soul, as retaining a piece of the original human. Dehumanization via HBOs would occur when people come to know that there are many thousands of HBOs being used and destroyed for medical research, and thereby come to think of all "human" life as ever so slightly existing to serve another's interest.

Transformation in Human Definition through Public Discourse

Even if this dehumanization effect were to occur, many people would undoubtedly conclude that the benefits of HBO and NC research would still be worth it. The argument would be that neurological disease creates its own form of dehumanization, so like so much else in life, HBO and NC research is a tradeoff. This is hard to argue with. However, there are probably ways of minimizing any possible associated dehumanizing effects.

How could creating NCs or HBOs result in a change in self-perception of humans so that we become slightly more animal or object-like, with possible attendant changes in our conception of how humans in general should be treated? How would the existence of a few thousand NCs or HBOs alter the collective definition of a human? Since the claim that a given scientific practice changes a society's culture is common in sociology and anthropology, it is to those fields we should look for an answer.

Consistent with the psychology literature on dehumanization, the claim is *not* that, after the first NC is created, everyone will begin to think of their neighbor as more like a mouse, or that, upon reading about HBOs, we will entertain the idea that a neighbor should be forced to become a research subject. Rather, the dehumanization resulting from any one technology is expected to be ever so slight, producing a change of, say, 1/100th of 1 percent in our conception of ourselves—if we assume for the moment that such an effect could even be measured. But although the effect would be tiny, it would be *cumulative*, with each technology slowly adding to the change, and the ultimate impact only really visible in retrospect.

How then do cultural ideas like this change? A basic sociological response is that culture changes through public discourse. Sticking with the simplest discourse mechanism, consider the labeling of entities. The vociferous debate about abortion shows anti-abortion advocates objecting to the words *fetus* and *embryo*, with their object-like overtones, in favor of *baby*, while prochoice advocates call for the inverse because of the theorized implications for the moral status of *baby*. Actual explanation is often unnecessary. The child's toy that says "cow goes moo" implies that humans are different from cows, without ever explicitly saying so.[18] A TV show from the 1960s about a hospital would show women as nurses and men as doctors without explicitly saying that men should be in authority over women.

Therefore, the act of creating an HBO or NC only matters to the extent that it influences discourse. In the extreme case, if an HBO or NC is created and nobody reads about it (like the tree falling in the forest that nobody hears), it has no effect at all. But, as I noted in Chapter 2, the science does not even need to occur to have a cultural effect. In principle, the technology does not even need to be possible—it only has to be debated. As Robert and Baylis state, "Indeed, asking—let alone answering—a question about the moral status of part-human interspecies hybrids and chimeras threatens the social fabric in untold ways."[19] Similarly, Leon Kass writes, in a commentary on selling human organs, that "We wonder whether freedom of contract regarding the body, leading to its being bought and sold, will continue to make corrosive inroads upon the kind of people we want to be and need to be if the uses of our freedom are not to lead to our willing dehumanization. . . . There is a danger in contemplating such a prospect—for if we come to think about ourselves like pork bellies, pork bellies we will become."[20] For Kass, as for Robert and Baylis, even contemplating the idea can lead to the cultural change. As the "Thomas theorem" in sociology from 1928 states, "If [people] define situations as real, they are real in their consequences."[21]

All else equal, a greater amount of a given discourse produces greater change.[22] We are all intuitively aware of this. For example, one of the many concerns about President Trump shared by his opponents was that by endlessly repeating claims of a stolen election, regardless of their credibility, he undermined faith in democratic processes.

Turning to the cases of NCs and HBOs, a debate in academic philosophy journals will have little effect because the public will not be aware of it. If the discourse escapes from academia through the media, there might be a small effect. But so far, the percentage of the public that has even heard of NCs or HBOs is very small. An ongoing stream of media coverage could spread the discourse; even the current limited media attention to the subject, with its language of "humanization" and the like, does seem to be promoting a weakening of the human-animal distinction. But this topic has not reached "water cooler" status, and a few newspaper articles are not going to change the culture. What would be required is for the knowledge of NCs and HBOs to spread, via television, movies, and social media, into the discourse that ordinary people pay attention to.

What would be further required for a cultural change of this magnitude is for NCs and HBOs—along with the awareness that these are human-animal mixes—to begin becoming part of everyday life. An example would be if NCs began being used widely in society in familiar roles such as pets or working animals. Another possibility would be if NCs became the incubators for growing HBOs for personalized neurological medicine. In any of these cases, these NCs would be extensively discussed.

For HBOs, it would take something even more fanciful, like constructing buildings ("farms") to house millions of HBOs engaged in some task for everyday life such as safety testing. While it is likely that HBOs will always remain behind closed laboratory doors, and thus unlikely to have any significant cultural impact, it is at least plausible that NCs could be found in the public realm within 30 years.

How to Protect the Foundational Distinctions and Avoid Dehumanization from NC and HBO Research

If violating foundational distinctions can lead to dehumanization, how can such distinctions be defended without forgoing important research? While it is inevitable that foundational distinctions exist, people will always be trying to add or remove particular ones.[23] Initial public repugnance toward an HBO or an NC caused by crossing a divide should not be slavishly deferred to; if foundational distinctions had never been challenged, all sorts of activities that are widely accepted today, such as homosexuality, would never have been accepted. As Leon Kass has written, "revulsion is not an argument," and "some of yesterday's repugnances are today calmly accepted." On the other hand, in crucial cases, "repugnance is the emotional expression of deep wisdom, beyond reason's power fully to articulate it."[24]

In other words, initial repugnance requires further inquiry rather than either dismissal or unquestioning acceptance. As one humanist writes, "if we cannot defend our revulsions by arguments that we ourselves can accept, we should instead work hard to change our attitudes of revulsion."[25] Put differently, using the anthropological language in which category violations are unclean, "rather than accepting the charge that expressions of disgust, monstrosity or unnaturalness are evidence of irrational fear and ignorance in public debates, we should actively look for what is being designated as 'dirt' by different speakers and different constituencies, and what kinds of purification rituals are being called into play."[26]

If we want to continue NC research without risking dehumanization, but also without dissolving the foundational distinction between humans and animals, we need to make sure the NCs do not invoke the foundational distinction. The same could be said about continuing HBO research while not damaging the human-object distinction.

Actually encountering an NC or HBO is not going to confront us with an unsettling violation of the foundational distinction, because it will not invoke the definition of a human that the public uses. However, when the public reads of NCs and HBOs, or more likely sees a YouTube video about them, the NC will be described as, say, a *human*-mouse chimera, and *HBO* will be spelled out as "*human* brain organoid." No matter what the public's visual perceptions might be, authoritative labels will be telling us that this is part human, thus confirming the violation of a foundational distinction.

We can see a solution to this problem by looking back to the anthropology literature on how to "clean up" a violation of a foundational distinction. Martijnje Smits identifies three approaches for addressing entities that violate foundational distinctions.[27] The first of these she labels "exorcism"—that is, forcing a radical halt in laboratory experimentation. This option, the most certain way to protect the foundational distinctions from NCs or HBOs, would require a ban on the technology. A kind of exorcism could also be accomplished by making sure that the public remains unaware of the technology. But exorcism is a last choice, since a ban would obviously foreclose any possible benefits of the advancing knowledge—in this case, treatments of human brain disease and dysfunction and for relieving the suffering of disease. And the continuation of research under conditions of secrecy would be undesirable because it is anti-democratic.

Her second approach is "assimilation," in which the category-crossing entity is used to change the public's foundational distinction. This is the hope of some bioethicists—that NCs could be used to teach the public that there should be no foundational distinction between humans and animals, with the incidental result that animals might be treated better in the future.[28] But I would hope to find a solution acceptable to those who want to retain the distinctions.

The third approach is "adaptation," in which the interpretation of the distinction-spanning entity is transformed to fit the existing distinction. To state the obvious, those who used to register disgust at IVF and heart transplants no longer do so, because people "normalize disgust."[29] It can be debated whether these specific normalizations happened through changing the meaning of the act or by destroying the foundational distinction, but in principle the description of an NC or an HBO can be changed to make it consistent with the foundational distinction.[30]

To adapt an entity like an NC or an HBO without challenging the foundational distinctions, we can be guided by a practice of the Nuer people of Southern Sudan. Mary Douglas writes that, among the Nuer, "when a monstrous birth occurs, the defining lines between humans and animals may be threatened. If a monstrous birth can be labeled an event of particular kind the categories can be restored. So the Nuer treat monstrous births as baby hippopotamuses, accidentally born to humans and, with this labeling, the appropriate action is clear. They gently lay them in the river where they belong."[31] In other words, a distinction can be protected by proper labeling.

Something similar already happens, probably unintentionally, in discourse about new technology. Maartje Schermer notes that, so far, the foundational distinction between humans and machines has not been threatened by mechanical implants in the human body, because of adaptation. Mechanical implants have so far been defined as "therapeutic," as repairing within the category of human. But, she wonders, what happens when an implant is not therapeutic? Presumably, adaptation will not be possible, and the machine enhancement will be seen to threaten the human-machine distinction.[32]

Specifically in the case of NCs, to protect the public's foundational distinctions we would start with not using the terms *humanization, humanized*, or *human hybrid*. Instead of referring to, for example, a "humanized mouse," perhaps we could call the animal a "Cebir" so as not to invoke the foundational distinction.

It is harder to avoid the word "human" when discussing HBOs. Certainly, those who want to avoid the perception that HBOs are human objects will avoid the term "mini-brain," which has been used in the media. By invoking the idea of a smaller version of a complete human brain, the term further equates the HBO with an actual human and thus breaks down the human-object foundational distinction. In order to clarify that HBOs are not human brains, somehow "part" needs to be included in the term or description.

Perhaps HBOs themselves could be given an invented name, such as "Torquors," just as we have been calling NCs "Cebirs" here.

The reader should be concerned that the labeling of scientific objects can have a rhetorical and persuasive function, where the term itself invokes a moral claim, which I consider problematic for accurate public debates. An example of the violation of this principle is "friendly mosquito," coined by the biotech company that created a mosquito that, when it breeds, produces mosquitos that die before reproductive age. This is clearly an attempt to use the label to connect to a moral framework.[33] Creating an equivalent term for NCs or HBOs would represent the manipulation of public debate and would thus be a violation of the norms of the public sphere.

Though these labels should be accurate descriptions of what is true about a technology, it is legitimate to avoid terms that have been linked in unrelated debates to particular moral conclusions. The renaming of "human cloning" as "somatic cell nuclear transfer" seems to have been an attempt to make the technology less controversial, but could be seen as avoiding irrelevant moral baggage. This is a borderline case. After the cloning of Dolly the sheep in 1997, "human cloning" came to mean producing a baby human that was a copy of an existing adult. This was arguably a distinct use of "cloning" from changing a somatic cell, and the baby-making version did come with a large moral apparatus attached to it. Perhaps in this case "somatic cell nuclear transfer" was justified.

For my case, a manipulative term would be one that tries to hide the fact that these entities are part human, and I may appear to be advocating such a manipulation above, since labeling an NC a "Cebir" avoids calling it human. But this would only be an attempt to be rhetorical and persuasive if we were using the philosophical or biological definitions of a human, in which a human possesses certain capacities. In those anthropologies, with humanity as a continuum, an NC is indeed "humanized," and this is why "humanized" was a natural label for the scientists and bioethicists who use these conceptions of a human. However, from the perspective of the definition of the human I found to be predominant among the public in my earlier study, the term "Cebir" is not rhetorical because an NC, no matter what its capacities, is not human.[34] Indeed, using the term "human" for an NC would itself be a case of scientist and bioethicist changing the anthropologies of the public.

Until a Cebir begins to act or move like a human—and, ultimately, until it is first born from a human—it is not legitimately a human but rather a kind of animal. Thus, from the public's perspective, using "Cebir" is not rhetorical. Using the name would simultaneously protect the foundational distinction between humans and animals and avoid the possible dehumanization that destroying such a distinction would seem to risk. Advocates of dissolving the foundational distinction between humans and animals by teaching the public a new anthropology should perhaps advocate for calling NCs "humanized mice" or "humanized monkeys," which would make plain that the distinction is being violated.

Avoiding the term "human" for an HBO may well be an attempt at persuasion. On the one hand, the term "human" that implies that an HBO could become equivalent to an actual human can legitimately be avoided, because an HBO can never obtain human status for the public. On the other hand, since it is true that HBOs are disembodied human parts that retain a connection to an actual human, it would be manipulative to hide that fact.

This could be a legitimate subject for scientific education. It would be wrong for participants in the public-policy bioethical debate to say that an HBO does not contain part of the soul of the cell donor, because the role of contributors to that debate is not to change the public's values but rather to reflect them. However, the public's willingness to believe in ephemeral connections seems to be somewhat based on the origin of the cells or the qualities of the HBO, and knowing the human origins of the stem cells that made the HBO could legitimately lessen belief in ephemeral connections.

Similarly, belief in ephemeral connections could be lessened, while not trying to dissuade the public in its views of souls or ephemeral connections, by educating the public about the lack of consciousness of HBOs. A connection between a non-conscious group of cells and the donor is presumably weaker than between a conscious group of cells and the donor. Presumably people would perceive a greater ephemeral connection between two consciousnesses. Therefore, if it is scientifically true that an HBO will not attain consciousness then teaching this to the public will lead to fewer people seeing ephemeral connections—without trying to change the major cultural distinctions that people hold.

The Ethical Limits of Technology and the Slippery-Slope Problem

To finish, I turn to ethical policy. At present, the HBO and NC technologies are in their infancy. The scientists who I have spoken with are hoping that limits will be imposed on them, restricting their use to something like "advancing human health." But could any such limits hold?

The public-policy bioethical debate is generally concerned with limits, as "ethical" implies drawing a line. For example, there is a 14-day rule for embryo experimentation, as well as, throughout most of Europe, a sharp line forbidding germline modification of humans. The bioethical debate about HBOs and NCs has vaguely asserted that experimentation should cease when an HBO or NC obtains "too much" consciousness. Though that level of vagueness is acceptable in an academic debate, much more specificity will be required if the insights are to be transformed into policy, since without the ability to define limits there can be no policy-based ethics. Here we need to consider possible ethical limits and, more importantly, which limits could actually be sustained in the face of advancing technological ability.

The best way to think about the limits for biotechnology is through a sociological version of the slippery-slope metaphor.[35] (I have described this type of analysis more elaborately elsewhere.[36]) The metaphor of the slippery slope starts with the most ethically meritorious position at the top and the most objectional position at the bottom. By merely stepping onto the slope at the top, it becomes more likely that we as a society will accept the currently objectionable position one step further down the slope. Having gotten to that step, it is then somewhat more likely that we will accept the next position down. Eventually we reach the bottom, where we never wanted to be when we took the first step. The ethically meritorious decision at the top changes the social conditions under which the next decision will be made, and so on as we slide down the slope. Though such slides are not inevitable, they become more likely under certain social conditions.

The classic sociological slippery-slope claim is that a society's decision to allow euthanasia for the terminally ill—considered to be a meritorious policy—changes how we view euthanasia in general. When euthanasia is next proposed for adults with severe mental illness, the concept of legal euthanasia is already familiar to us and thus more likely to be agreed to. After institutionalizing euthanasia for mental illness, a proposal to expand the practice to mentally ill children is made. By that point, we would presumably be used to broadening applications of euthanasia and there would be procedures and institutions in place—for example, euthanasia clinics—to make the next step down the slope even easier.

Eventually, we would reach the bottom of the slope, perhaps defined as offering euthanasia to anyone who just does not want to live anymore. When we took the first step, we assured ourselves that we would never reach the bottom—but here we are. Again, the first step does not automatically lead to eventually descending to the bottom, but merely increases the chances of such a slide on this slippery slope.

To start a slippery-slope analysis, we have to define the bottom (while looking down from the top). In the human gene-editing debate, the bottom is perhaps a world where all human reproduction is technologically mediated in order to produce humans designed to specification, which could result in a radically stratified society.[37] In the world of artificial intelligence, the bottom might be allowing computers to obtain so much power that they could decide humans are superfluous. For euthanasia, the bottom would perhaps be deciding that certain people should be put to death against their will because they are simply a drag on the collective.

I will conduct a slippery-slope analysis for HBOs and NCs using the views of both the public and the public-policy bioethical debate. For bioethicists and scientists, whose primary concern is consciousness, the bottom might be trapping an HBO with human consciousness in a dish, torturing an NC through a laboratory experiment, or endowing an NC animal with human-level consciousness.

Using the public's anthropology, the bottom for the HBO debate might be represented by an HBO that clearly retains the soul or essence of the person from whom it was grown, as indicated by its acquiring the social behaviors used to define humanity—for example, sociability, responsibility, and the ability to fall in love. The true bottom would be when the donor and the HBO had clearly become the same self.

I think we can say that no HBO will ever reach that stage, but we could get partway there by achieving something like an HBO capable of controlling a mechanical device. Envision here the robot with the brain in the vat. Science fiction can be useful in assisting a society to imagine what ends it should and should not pursue, and a technology does not need to be possible for us to morally reason with it.

As for NCs, the bottom of the slope, in the public's view, might be an entity that perfectly mixed the core elements of an animal and a human so as to maximally confuse the categories. If I were to ask the public, they might say "I am not worried about the mouse in the lab with a 3mm piece of human brain in its skull, but I am worried that you will eventually create a talking mouse." That is, the bottom is represented by Mr. Ed or the Planet of the Apes—the stuff of science fiction. People will want a limit far above this bottom of the slope.

How Do We Slip Down the Slope?

The slipperiness of the slope depends on the terrain, and the terrain is defined by the moral concern about each position on the slope. Metaphorically, the slope has a number of possible paths down-hill, each defined by a distinct moral concern or moral value. So, to identify the terrain on a particular path on the slope, we would ask: What is morally relevant about putting a 4mm piece of human brain tissue into a mouse? The slope's terrain for bioethicists, their path, would at each step concern animal abilities or capacities, and would range from "the capacity of a glimmer of consciousness" near the top of the slope and the capacities required to go to college at the bottom. The terrain for the public would be the human-animal divide: the extent to which the NC appears both human and animal. (I will use the NC for my examples because it has a longer history and has more established discourses associated with it.) At the top of the slope is the mouse that looks and behaves like a mouse; near the bottom is Mr. Ed, the animal who behaves like a human.

In the absence of a sharp line between cases, a debate would be likely to slip one unit down the slope.[38] There are two features of the terrain that would tend to remove the sharp line and allow for the slipping down the slope on a moral path. The first is "continuity vagueness," where the terrain has a continuous measure.[39] For ex-ample, if the terrain for a euthanasia slope, the moral concern, was proximity to time of death due to disease, euthanasia could be used if the person had 30 days left to live. This could easily become 31 or

32 days, because there is no identifiable moral distinction between 30, 31, and 32. Similarly, to foreshadow, what is the difference between a 4m HBO and a 5mm HBO? Nothing, as we cannot construct a moral distinction between a 4 and 5mm HBO.[40]

The second feature of the terrain is "similarity vagueness," where "measurement is not possible, irrelevant, or when it depends, at least in part, on imprecise components."[41] Think, for example, of trying to make a moral distinction between a mouse and a chipmunk. Probably the most famous expression of similarity vagueness was that of Supreme Court Justice Potter Stewart who, when asked to define obscenity, said "I know it when I see it"—but that is a limit that cannot be enacted into policy.

It is, then, the similarity between two adjacent locations on the slope that allows the debate about HBOs or NCs to slip downslope. For example, creating an NC mouse and creating an NC chimpanzee are similar, and the debate can slip between the two points on the downhill path when the only moral concern is value for human health, because this moral value is not concerned with the nature of animals. The only standard would be, "How much does this advance knowledge of human disease?" If it is the same, creating both NCs are justified, and the debate slips to the chimpanzee location.

If the moral path leading down the slope is concerned with the consciousness of life forms, then this creates the distinction between mice and chimpanzees. Thus, we could cite the difference between mice and chimpanzees as a basis for the ethical distinction.

The final image for my slippery-slope metaphor is the barrier. Bioethical debate can largely be described as the setting and justifying of barriers. As Granville Williams wrote many years ago, "all moral questions involve the drawing of a line."[42] Philosopher John Harris writes that "slopes are only slippery if they catch us unawares and we have strayed onto them inadequately equipped." According to two legal theorists, "slippery slopes . . . can sometimes be resisted by standing on easily enforceable bright-line rules."[43]

A strong barrier (a limit) then separates two steps on a path leading down the slope that cannot be made similar through similarity or continuity vagueness, with any action above the barrier being morally acceptable and any action below unacceptable. While ethical policy does not use the term "barrier," that is essentially what such policy entails: Before this point, the actions are acceptable; below this point, they are not. What discrete acts of creating an HBO or a NC can be made so dissimilar that they can serve as barriers in the debate? That will be the question for the remainder of this chapter.

For a barrier to hold, the acts on either side of the barrier have to be perceived to be utterly different, incapable of being merged through continuity or similarity vagueness. However, a value change in the society can create a new path down, bypassing established barriers by using values that make the two sides of the barrier the same.

As an example, let us consider one of the strongest barriers in any debate in bioethics history, the barrier between *somatic* and *germline* human gene editing. Somatic gene editing is editing that affects only an existing human body, whereas germline editing will affect a living person's descendants. Somatic HGE has traditionally been regarded as above the barrier and thus acceptable; germline HGE has been below and thus not acceptable. The distinction between the two is not continuous, so continuity vagueness is not an issue. It also is not subject to similarity vagueness, because the two are clearly different in nearly every way: one applies to the present, the other to the future; one to the person, the other to the species; one is about the self, the other is about someone else; and so on.

This barrier held for decades on the old moral path leading down the slope. However, American values changed and a new path opened, defined by the value of individual freedom. By this value, somatic and germline are not different if people can autonomously choose to engage in either technology. For example, people would have the reproductive freedom to have germline modified children. There is no germline barrier on this path down the slope, and scientists are now planning germline modifications.[44]

In sum, the barrier is the response to questions from the public as to why allowing the NC mouse today will not lead to the human-behaving animal in the future that violates the public's values. In future years technology will advance and various barriers will be proposed, and it is important to estimate their strength beforehand. It is important to ask whether pushing through what might turn out to be the *only* viable barrier between the present and the bottom of the slope is really worth it.

Possible Barriers on the HBO Slope

Before we begin examining possible barriers on the HBO slope, let me dispose of one "barrier" that people often like to imagine on a bioethical slope: "scientific reality." The claim is that it will always be impossible to do X, and therefore X can serve as a barrier. For example, we could say that it will always be impossible to endow primates with human behaviors or appearance, and therefore there is no need to worry about the slope below that point. Unfortunately, what we think of as impossible today has the tendency to become possible. HBOs are a case in point: If I had written in the 1980s that we will grow parts of human brains in dishes from cells taken from someone's skin and that serious people would be concerned that these could obtain consciousness, I would have been considered a science-fiction writer.

Obviously, a slippery-slope analysis of the HBO debate will be almost purely speculative, since the technology and ethical debate are so new. That said, an unfortunate aspect of the debate about HBOs is that I see few viable barriers. We can see how the top of the slope is totally acceptable and the bottom totally unacceptable, but there is no barrier that can be created in-between.

Let us start with likely scientific developments for which a limit would need to be placed. One development is that we can expect

HBOs to get larger. A 4mm HBO will never have any function that we associate with a human, but possibly a 200mm HBO could. But this is a textbook case of a continuity vagueness problem. If we set a barrier at 4mm, why not 5mm, or 6mm, and so on? If we express size by weight or number of neurons, obviously we have the same problem. There is no value in the debate that can be used to make a moral distinction—a barrier—between a 4mm and 5mm HBO, a 20mm and 21mm HBO, or a 199mm and 200mm HBO. A 200mm HBO might elicit an "Eeewww!" reaction today, but by the time 199mm HBOs in dishes have become normal, nobody would notice the 200mm version. A similar slippage involving the 14-day rule on human embryo experimentation is currently underway, with scientists challenging the time limit as arbitrary.[45]

A barrier could perhaps be anchored by equating the size of the HBO and the brain size of various animals—say, at the brain size of the average dog. But this would then require a moral distinction between animals with different brain sizes—between a mouse and a rat, or a rat and a pig, which is not possible with the values I am aware of in this debate. This also implies that our moral rankings of animals are tightly correlated with brain size, and I suspect the correlation is not very tight. A barrier between HBOs at different sizes (and size analogs, such as numbers of cells) is unlikely, because such a barrier would not be structurally sound.

A similar barrier that would not have continuity or similarity vagueness could be established at communication between brain parts. What we think of as human abilities requires communication between the parts of the brain, and a barrier could be built at the point where communication begins. But this assumes that brain scientists have reached consensus on what constitute different parts of the brain and what communication means. Such a barrier would be very high on the slope and would definitely stop us from getting close to the bottom, because tissue from just one part of a brain is arguably not itself a "brain." However, this concept ultimately turns out not to be very strong, because it is hard to imagine the moral value that could be used to make a moral distinction between a piece of a brain and two connected pieces of a brain. Moreover, it may already be too late, scientists have begun creating rudimentary connections between HBOs representing different parts of the brain.[46] Having reached the point at which researchers are already engrossed in their task, it becomes very difficult to call a halt to such research.

A conceivable barrier could be located at the distinction between creating and manufacturing—between treating HBOs as quasi-sacred objects, on the one hand, and considering them as objects to be manufactured, on the other. But while the earlier chapters suggest that the value underlying this distinction would be supported by the public, any barrier we could imagine would actually be hopelessly subject to continuity and similarity vagueness. How many HBOs does it take to be engaged in manufacture? 50? 50,000? How would we tell the difference between creating and manufacturing were we to see it?

If nature would cooperate, a possible barrier would be between "conscious" and "not conscious," with anything above the barrier being not conscious and anything below the barrier conscious. The terrain on this path is defined by concern about harm, and it is only conscious entities we are worried about harming. Unfortunately, the level of our knowledge makes the placement of this barrier impossible. The first problem is that we cannot even define what consciousness *is*. Our ability to distinguish no consciousness (a fruit fly?) from some consciousness (a bird?) remains extremely speculative,[47] and such a barrier would be hopelessly subject to continuity and similarity vagueness.

Another hypothetical barrier would be at "human-level consciousness." However, our ability to place the barrier on the slope would be problematic, since, again, we lack the ability to precisely define the consciousness of even an ordinary human, to say nothing of the defining variation in consciousness between humans. Moreover, this barrier would be really far down the slope, and many people would oppose creating even HBOs with much less consciousness than a human. Levels of consciousness simply cannot be used to establish a barrier that will be strong enough to hold once scientific advances began pushing against it.

Though there are a number of possible barriers that could be structurally sound, in that the acts on either side are recognizably distinct, these barriers are not on any moral path that concerns people, and therefore would never be built. One might be at vascularization, which would feed nutrients to the HBO. HBOs without vascularization would be above the barrier and those with vascularization below. Whether or not an HBO was vascularized would be clear, so this barrier would not be subject to continuity or similarity vagueness. But what value supports the moral difference between vascularization and non-vascularization? If the public's definition of a human included flowing blood, this could work, but that is not the case, and if your ethical value is based on relieving the suffering from human disease, vascularization is irrelevant. People who want to vascularize an HBO will say that this barrier is not morally relevant, and they would be right.

There is one barrier that would be weak but possible and would fit the values of bioethicists and the public. A barrier could be placed at giving an HBO experiences, with "no experiences" upslope and acceptable and "given experiences" downslope. A human brain cannot develop and wire itself together without experiences, and consciousness is not possible without experiences. For the public's version of a slope, the public's definition of a human includes interaction, so a non-experienced HBO would definitely never be human. An HBO with experiences would presumably trigger more belief in ephemeral connections, so the barrier would be supported by the public's moral distinction between humans and objects.

The experience barrier does not appear to suffer from similarity or continuity vagueness: either you shine a light on the HBO or not; either you hook it up to a robot or not. This barrier would be set very high up the slope. However, scientists are already giving HBOs rudimentary experiences, so it seems we have already passed that point.

It is possible that scientists and bioethicists could install a barrier that is susceptible to a slippery-slope process in order to signal that they intend to stop before the bottom of the slope, but without actually stopping. This is essentially what the 14-day rule for human embryo experimentation is, and it was possible because scientists were unable or did not want to move beyond it at the time. But soon after scientists are able to or want to move beyond the 14-day limit—effectively, right now—this barrier will inevitably be knocked over. The 14-day rule is thus less a barrier than a speed bump.

Perhaps we could instead propose a size limit for HBOs—for example, 20mm—that is currently far beyond scientists' ability to achieve but still far smaller than would be necessary to create a fully conscious HBO. But when scientists are able to achieve an HBO of that size, it will be argued that the "20mm rule" is not actually grounded in any moral value and should be removed, and the slide would continue. A slippery-slope analysis could tell us which proposed barriers are actually nothing but speed bumps.

The HBO debate exemplifies the utility of slippery-slope analysis. No one wants to reach the bottom of the slope, but there is no unbreachable barrier that can prevent it. It therefore seems likely that this debate will continue down the slope at the rate at which our technological abilities progress. Those who fear this will be left with the hope that biological reality will one day reach its limit at "humanly impossible."

Possible Barriers on the NC Slope

It would be accurate to describe the NC debate as the part of a broader slippery slope concerned with mixing humans and animals. We can gain clarity by examining some barriers on this slope that have already fallen. The first was the first physical mixing of a human and an animal, which was a potentially strong barrier. This was bypassed when the moral value of relieving the suffering of disease became the most powerful value in these debates—the widest path down the slope. As described in the first chapter, this first step supposedly occurred in 1501 when a bone from a dog was put into a human, and was followed in 1667 when a man was infused with lamb's blood.[48] Similar attempts between the seventeenth and nineteenth centuries were few and not of obvious efficacy, but by the late nineteenth century medicine had become more ambitious, and in 1893 a British doctor transplanted parts of a sheep pancreas into a 15-year-old diabetic child. And in the 1920s 43 men received slices of chimpanzee testicles.[49]

Today, implanting heart valves from cows and pigs is standard medical practice, and scientists are continuing to work on genetically modifying pigs to make their organs more compatible with humans.[50] Five centuries of attempts have thus led to the first few steps down the slope.

The public seems to have largely accepted xenotransplantation from an animal into a human. While it does violate the human-animal distinction in a superficial way, and thus undoubtedly once evoked a "yuck" response, xenotransplantation does not threaten the public's definition of a human. Xenotransplants do not change the humans' appearance and certainly do not alter the fact that the humans were born of other humans, nor do they affect the social interactional behaviors that are part of the public's definition of being human. More importantly, the direction of these transplants preserves the moral superiority of humans, in that it is the animals that are being sacrificed to save the humans.

The Human-to-Animal Barrier

This points us to a possible barrier lower on the slope that would not be susceptible to continuity or similarity vagueness: that is, animal-to-human transplants, which would lie above the barrier and thus be acceptable, as opposed to human-to-animal transplants, which would lie below the barrier. Because the difference between a human and a nonhuman animal is categorical and unambiguous, this barrier would be very strong. It would also tightly fit with the value of retaining human superiority over animals, in that animals would be sacrificed for human health in the same way they are already sacrificed for food and medical research.

This would seem to be a perfect barrier. However, the primary value lurking in this debate has already changed the moral terrain, so that the acts on both sides turn out to be moral equivalents. The value of beneficence (toward humans) makes animal-to-human and human-to-animal transplants similar, since they both promote human health. If beneficence is the primary value, the barrier then becomes moot.

The animal-to-human barrier fell in the late 1980s with the "humanized mouse" containing human genes or cells. In 1988 a mouse genetically modified to have a weakened immune system was engrafted with human fetal tissues, adult blood cells, and hematopoietic stem cells.[51] As technology has continued to develop, many mice have been genetically modified to give them human diseases and conditions such as muscular dystrophy. Experiments such as those described in the recent article "Mice with Human Livers," to take one example, allow for the study of human disease without having to experiment on humans.[52] For at least 30 years, many animals have received human tissue; however, notably, none of these experiments have significantly impacted the public's definition of a human, because the human component involved (e.g., the liver) was not central to human identity, and thus have not been felt to violate the foundational distinction between humans and animals.

Barrier at Animal X

Now is a good time to rule out the initially plausible-sounding barrier at a particular animal. Even if we could construct a continuum of animals ranked by their capacities, with worms near the top of the slope and chimpanzees near the bottom, any barrier would suffer from extreme continuity vagueness on whatever the difficult-to-describe scale would be. If we posit that the standard lab mouse would be rated at 20 out of 100 on the scale, there would be another animal at 21, though the difference between 20 and 21 would be not only arbitrary but probably unmeasurable. Something similar could perhaps be said for any proxy for animal capacities, such as skull size, which would limit the size of the brain and presumably limit its capabilities. Regardless, research would move progressively up the scale to more advanced animals as we slipped down the slope.

There could be a slightly stronger barrier at branches of the evolutionary tree, because these distinctions, while themselves arbitrary, have taken on a fact-like status. If we assume that research will continue on rodents, which are fairly distantly related to humans among the mammals, we could establish a boundary at primates. Such a consensus would at least be structurally sound and resistant to continuity and similarity vagueness.

The problem is that this barrier does not map onto the public's values. To make a simple point, the public probably holds dogs in greater esteem than squirrel monkeys. Thus, it seems more persuasive that a barrier could be erected not at primates but at the great apes—orangutans, gorillas, bonobos, and chimpanzees. While I see no distinction between these and other primates that would withstand continuity and similarity vagueness, the great apes have already been put in a different category by many legal systems in the world.[53] Such a legally enforced barrier presumably has some durability; nevertheless, the law could end up being changed.

Barrier at the Brain

We now return to the focus of this book. The brain—the neuro—did represent a structurally sound barrier, in that all *neuro*-chimeras would be below the barrier, whereas all non-neuro chimeras would be above the barrier. The brain is distinct from other parts of the body, and there is no vagueness in the difference between the mouse brain and the mouse leg.

The initial moves beyond this barrier were small steps down the slope involving tiny amounts of human cells or tissue in animals that few people care about (e.g., mice), in ways that did not create any markers of humanness beyond the mere fact that human cells were involved. As almost always when a barrier is breached, the public reaction was not horror but rather indifference to the act itself, because it was but one small step beyond what was already upslope. The steps are slippery because they are small.

The motivation for writing this book is that this barrier has been knocked over. While half of the public would support keeping the barrier, this is probably because it is the only strong barrier above the point on the slope where their values would actually direct them to stop. Using the public's definition of a human would require a barrier not at the brain, but at brain tissue that leads to markers of humanness that in turn invoke the human-animal distinction. This ideal barrier would be subject to the two types of vagueness, so people settle for a defendable barrier above.

The Barrier at Human Behaviors

Since the public's primary value is preserving the foundational distinction between humans and animals—from the perspective of the public's definition of a human—a barrier exactly at human behavior would fit perfectly with that value. However, structurally this is very weak because it suffers from extremely strong continuity and/or similarity vagueness. A basic rule in slippery-slope analysis is that if you cannot even define it, you certainly cannot build a barrier upon it, and "humanlike" behavior probably cannot be precisely defined.

The Barrier at Medical Research

Another barrier that invokes the value of beneficence prevalent in bioethical debates would be one that divides medical from other uses of NCs. Connecting to such a dominant value would be very powerful. At present, all the human-to-animal chimeras are used for medical research; that is, they are justified by the value of generating knowledge that will lead to the beneficent reduction in human suffering due to disease. This medical-research barrier is implicitly supported by most HBO and NC researchers.

However, if the technology became more advanced, we can imagine others wanting to use NCs for military or industrial purposes, which would be below a medical research barrier. For example, in Robert Heinlein's science-fiction short story "Jerry Was a Man," a strain of humanized apes has been created to do labor that humans did not want to do.[54] Given that humans already breed animals for tasks that we do not want to do (e.g., the donkey), it is inevitable that people would consider this application. A medical research barrier could also be installed in the HBO debate, with using HBOs for industrial purposes or for art being below the barrier.

This barrier could hold if people defined the value of "beneficence" as only referring to healing disease in human bodies. But the concept of "disease" is somewhat vague, and the first breaching of the barrier might be the beneficent creation of NCs that do not directly treat disease but that help people who have diseases, such as developing more effective guide dogs for the blind. From there it would be a short slide to the creation of NCs that can perform tasks that could threaten the health of humans, like working with toxic materials. These would all be considered to be about health. If the technology gets to this point, this barrier will not hold.

The Barrier at the Germline

One strong barrier remains. Recall that, to paraphrase the famous aphorism about the tree falling in the forest, if the NC mouse is created in a lab and nobody hears of it, it will not have any impact on social conceptions of humanity. Therefore, the raw number of NC animals will strongly influence the impact on public opinion. Above the barrier would be the creation of any NC that is created somatically, presumably through surgically implanting an HBO or through modifying an animal blastocyst. Each NC would have to be created individually, which would mean that these foundational distinction-threatening entities would remain rare and locked in labs. For NC animals to actually enter the public discourse would probably require creating a species of reproducing animals, analogous to the genetically modified strains of reproducing chimeric (non-neuro) mice that are already in laboratory use.

This barrier at the somatic/germline distinction is structurally sound, in that the distinction is not subject to continuity or similarity vagueness, and there is no ambiguity about modifying the genes in the reproductive cells of an animal.

This barrier would also be supported by values. My earlier analysis shows a distinction between preservationism and conservationism, between believing that nature should exist "as is" and that humans may modify nature for their own ends. That is, creating nonreproducing NCs would lie above the barrier, and creating a new species of animal would lie below. Activity above the barrier would be seen as fiddling with nature as we have always done, but activity below would be inventing species that God or nature had never created. This value would presumably be strong

enough to overcome what would be the value of economic effi-
ciency (since constantly creating animals is inefficient). My analysis
also showed that this value is not universally held by the public, and
it is not clear which subset of the public would be considered in any
ethical debate. If it turns out that medical research is advanced by a
subspecies of NC animals, the barrier may not survive reliance on
beneficence as a value.

Conclusion

This book has examined the ethical debate about HBOs and NCs.
Using as my foil the public-policy bioethical debate that influences
government policy, I have shown how the actual public sees the
moral issues differently. In thinking about HBOs and NCs, the
public is not very concerned about consciousness. In the case of
HBOs, the public tends to reject the materialist reductionism that
body and mind are ultimately nothing but chemicals, and tends to
see humans as having something like spirit, which produces beliefs
in various ephemeral connections between a human and an HBO.

One of the primary values also moving the public is the foundational distinction between humans and animals, a distinction
that NCs violate. According to a many-decades-old humanistic
concern, once the foundational distinction is breached (by NCs or
other entities), we will treat each other slightly more like we treat
animals—that is, poorly. For those who fear that the existence of
NCs will breach the foundational distinction, I have suggested
some remedies.

The public bioethical debate that influences policy should in
principle include the public's views. It should also be about limits.
Bioethical debate was never intended to be the public-relations
arm of science, but was instead meant to be an intermediary for
incorporating the public's values. In this final section, I have used

the slippery-slope metaphor to propose various possible limits on biomedical technology and to predict the odds of each of them holding. All of this is premised on the assumption that the public will at some point begin to pay attention to the HBO and NC debates. I hope that, at minimum, I have provided the reader with an intellectual basis for continuing to engage with these issues.

Methodological Appendix

Survey Sampling

My survey data were gathered in three stages. (Because of the nature of the three empirical projects, they were all deemed by my university's IRB to be exempt from oversight.) First, cognitive testing was carried out on 35 respondents, as described in Chapter 3, endnote 3. Second, as a further test of the survey questions, I surveyed 497 respondents on Mechanical Turk. The resulting data led to a few small improvements in the flow and wording of the survey, and the Mechanical Turk data were not further analyzed.

Third, having created the survey instrument in Qualtrics, I then contracted with the online survey firm Lucid to provide the respondents for the final data. Lucid has been extensively used in social science, particularly by political scientists,[1] who conduct ongoing studies to ensure that it remains a viable source. For example, tests have concluded that the data gathered in the Covid era have been as legitimate as data gathered previously.[2]

Lucid is an opt-in poll, and research in the past decade has shown "few or no significant differences between traditional modes [of survey administration] and opt-in online survey approaches."[3] Lucid sets quotas for age, gender, race/ethnicity, and region to reflect US Census figures. Additional post hoc weighting adjustments to ensure that the sample represents the US population are described below.

I surveyed a national sample of 4,619 US adults, with a preliminary wave in early July and the remainder in early August 2021. In the invitation to participate, the survey was described as an investigation of "attitudes toward biotechnology." The informed-consent page stated that "The purpose of this research study is to determine what the public thinks about developments in biotechnology and its application to medicine." Therefore, people who were not interested in these topics were less likely to be in the sample. This bias was seen as acceptable because all surveys are similarly biased toward including those interested in the topic, and any future activity involving these issues in the public sphere will involve people who share this interest.

Attention

One of the primary challenges for any survey is ensuring that the respondents are paying attention, because inattentive respondents produce less accurate survey results. By excluding the inattentive, I bias my results toward the attentive. This is an acceptable bias because I suspect it is respondents like this who are more likely to engage in debates on this topic in the public sphere.

In the years before online surveys were possible, it was difficult to say what percentage of respondents to a face to face survey, a phone survey or a mailed pencil-and-paper survey were paying attention. We can presume that the inattentive excluded themselves from gold-standard face to face surveys of previous decades by refusing participation. Part of the justification for using opt-in samples is that they have a low cost per respondent, so the analyst can buy more responses and only analyze those from respondents who are truly focusing on the survey. In recent years, social scientists have developed techniques to screen out the inattentive from surveys.

One method for identifying the attentive is to use "attention check" questions, questions whose answers are obvious if the respondent actually reads the instructions.[4] In my survey, after the first few questions, I asked a slightly modified version of the "What is your favorite color" question developed by Adam Berinsky.[5] In Lucid, upon failing the attention check, such respondents are dropped from the survey. In my case, 24.8 percent of those who started the survey failed the attention check and were removed, which is a fairly standard rate.[6] I also removed a few who made it past the attention check but ceased responding to the survey after the first vignette, for an overall exclusion rate of 27.1 percent.

The demographic data provided by Lucid shows that certain groups were more likely than others to fail the attention check or be removed for non-completion: nonwhite compared to white (32.9 percent vs. 23.9) and males compared to females (28.2 vs. 25.9). Those who failed were also younger (on average, 43.4 years for those who failed vs. 46.2 years for those who passed) and less educated (on a 6-point scale, 3.3 for those who failed vs. 3.7 for those who passed). Again, Lucid removed the respondents who failed before they reached the subsequent question. I later weighted its data to account for these demographic biases.

An obvious response to lack of attention is to ask whether the inattentive can be induced to pay attention—in the evocative title of a paper on the topic by Berinsky and colleagues, "Can We Turn Shirkers into Workers?"[7] Berinsky's group attempted to induce attention through methods such as not allowing respondents to move on in the survey until they had answered the attention-check question correctly, or warning respondents that their answers would be checked carefully. These methods did induce attention to the attention-check question but not to the rest of the survey.

Diana Mutz suggests improving attentiveness in a vignette survey by asking a question about it in the middle of the vignette. Inspired by Mutz's suggestion, at the end of each of the vignette texts but on the same page of the survey, I asked a question about the vignette prefaced by "Now we want to see which parts of that description we were the most clear about. Please look at the question and then go back and re-read to find the correct answer."[8] Though I could not remove respondents based on their eventually answering this question correctly, because different factor combinations could have made the question easier or harder to answer, unbalancing the experimental design, I posed it simply to get the respondent to pay more attention by rereading the vignette.

Another method of ensuring that only attentive respondents are included is to remove those who answer a survey so quickly that they could not have been paying adequate attention. Since there is no set standard for "too fast," I removed the fastest 20 percent of respondents, those who finished in 454 seconds or less. And while other studies winnow out the inattentive by asking multiple attention-check questions, I myself winnowed by using a single attention-check question, along with the survey-completion time. [9]

Scholars have also shown that a respondent's attention may fluctuate during a survey. In my own survey, it may be that certain respondents just skimmed the vignette, the part of the survey that actually required the most attention. [10] Excluding these respondents for being inattentive was also important,[11] so I also excluded those who raced through the vignettes.

While I could have created a separate time threshold for each of the three vignettes in the survey, for the sake of simplicity I removed the consistent racers, those who, while having passed the overall time threshold, were among the 25 percent who spent the least time (between the moment of the screen opening and moving to the next screen) on all three of the vignettes—an average of 37.4 seconds for the gene drive mosquitos vignette (not analyzed in this book), 39.1 for the HBO vignette, and 24.8 for the NC vignette. As in the basic attention-check question, those excluded from the sample for going too fast—either in general or only on the vignettes—were disproportionately young, male, nonwhite, and less educated. Since the different treatment conditions within each vignette have nearly the exact same number of words, excluding on the basis of time was not expected to interfere with random assignment.[12] (Time spent on each vignette was uncorrelated with any factor, and the exclusion threshold was based on a combination of all the vignettes, so this exclusion did not unbalance the random assignment of the vignettes.) After these deletions, 2,095 responses remained for analysis.

Weighting

It is important, post hoc, to weight the data so that it better represents a random sample of the population. Given the exclusion of some demographic groups on the attention-check methods, the weighting needed to incorporate education level, gender, race, and age. To adjust to the population parameters, it was necessary to know what the true population parameters were. For demographic measures I relied upon the US census and other government surveys.[13]

Samples should always be weighted on variables that are theorized to matter to the phenomenon in question. Studies have shown that, when adjusting opt-in samples to population benchmarks, basic demographics are typically not the most important consideration; instead, it tends to be people with particular political and religious identities who are most often under-enrolled in these surveys.[14] For example, political scientists weight these types of samples to make sure there are the right numbers of Democrats and Republicans.[15]

Regarding public views of biotechnology, religion and political ideology are the most important factors for which we have benchmarks from the population. Again, the challenge is to obtain a measure of each group's true representation in the population. For basic demographics, we can presume that the US census is true enough. For party identification and religion, I relied upon the 2018 General Social Survey (GSS), the gold-standard social survey in sociology. Since the GSS measures religion more extensively than I can do with my survey, I created a simplified weighting.

In the 2018 GSS I identified Catholics, non-literalist Protestants, and literalist Protestants, with the latter two distinguished by their response on the view of the Bible question.[16] The 2018 GSS identified 21.3 percent of the population as Catholics, 23.5 percent as non-literalist Protestants, and 22.7 percent as literalist Protestants. In my own survey, using the same questions, and before weighting, 18.4 percent of those who passed the attention checks were Catholics, 22 percent were non-literalist Protestants, and 16.5 percent were literalist Protestants.

For party identification, I used the well-known party-identification questions from the GSS in my survey. Political party and religion are increasingly correlated in the United States, so for these two variables I created appropriate categories that crossed the two. Thus, I weighted to ensure the correct number of Catholic strong Democrats, evangelical strong Democrats, and so on. In order to avoid sparse cells, I collapsed the "not strong [Republican/Democrat]" and "independent, near [Republican/Democrat]" into a single category, for a total of twenty religion/party ID categories.

For weighting, I used the Stata svycal calibration rake command.[17] One concern with weighting is that some cases may have overly large or small weights, and a single idiosyncratic case may thereby end up having too much influence on the results. Since there is no objective method of selecting where to trim, this decision has to balance bias and error.[18] Common bounds in social science are weights of .2 and 5 or .125 and 8. When initially calculated, about 1 percent of respondents had a weight above 5, and 1.3 percent had a weight below .2. I therefore trimmed to .2 and 5.

Creating Religion Variables from the Survey

My survey asked a set of religious-identity questions. The first was, "What is your present religion, if any? If more than one, click the one that best describes you." My choices (closely based on the work of Elaine Ecklund and Christopher Scheitle) were Protestant, Catholic, "Just a Christian," Jewish, Mormon, Muslim, Eastern Orthodox, Buddhist, Hindu, "Nothing in particular," Agnostic, Atheist, and "Something else" (with a write-in box).[19] As expected, far too few respondents selected Jewish, Mormon, Muslim, Eastern Orthodox, Buddhist, or Hindu for separate analysis, so I combined these into an uninterpreted "Other religion" dichotomous variable. This was included in models to produce the proper comparison.

"Something else" was selected by 6.3 percent of the respondents; these were asked to fill in a box, and the responses were hand-coded. The majority of these were also assigned to "Other," typically because they referenced even smaller religious groups like Jainism. Some cases that I describe below were recoded into other dichotomous variables.

Those who selected "Catholic" were coded as Catholics. Those who selected "Nothing in particular," "Agnostic," or "Atheist" were assigned a nonreligious dichotomous variable, as were the respondents who expressed nonreligion (e.g., "None") when providing supplemental description for the "Other" category. Following the self-reported identification measurement strategy used in the sociology of religion, those who selected "Protestant" or "Just a Christian" were given an additional identity choice, "Which of the following terms best describes your religious identity,"[20] with the options "Fundamentalist," "Conservative Protestant," "Evangelical," "Born-again Christian," "Mainline Protestant," "Liberal Protestant," and "None of the above." The first four were assigned a "Conservative Protestant" dichotomous variable, and mainline and liberal Protestants were assigned to the "Liberal Protestant" dichotomous variable.

My survey also included a measure of certainty of belief in God, and (for Protestants) belief in the status of the Bible, as well as another question that asked respondents how religious they were. These additional measures were used to make the proper identification of the respondent's religious identity.

The few conservative Protestants who also claimed on the biblical-exegesis question that the Bible was human-made, and also chose "I don't believe in God," "I don't know if there is a God," or "I don't believe in a personal God but I do believe in a higher power of some kind" were moved to the "Liberal Protestant" identity, since these statements are incompatible with conservative Protestantism.

It is a growing challenge for surveys in American religion that Americans are decreasingly identifying with denominations or with generic religious-identity labels.[21] That is, many respondents whom academics would classify as belonging to a particular Protestant tradition either do not know they are Protestants and do not recognize that label in a survey, do not recognize or reject any of the identity labels used in the Protestant community, do not know if their church is a member of a particular denomination, and/or reject any identity beyond "Christian."[22]

In this sample, 19.8 percent of those asked the specific Protestant-identity question selected "None of the above." A good portion of these respondents do not recognize or use the term "Protestant." This is demonstrated by the fact that, of those asked this follow-up question, 34.1 percent were asked because on the previous question they had selected "Just a Christian," and 33.0 percent of the latter ultimately rejected a label in the follow-up question.

I therefore created a dichotomous variable for these identity-rejecting Protestants, and also included the respondents who used a term that was clearly Protestant (e.g., "Adventist" or "Baptist") in the "Other" religion text box. Moreover, in the same way that I used belief about biblical exegesis and beliefs about God to further separate conservative Protestants, I took the same group of skeptics from the identity-rejecting Protestant group and put them in the "Low-commitment Christian" group (while recognizing that many identity-rejecting Protestants are nonetheless quite involved with religion and are essentially a different type of Protestant).

The general expectation among sociologists of religion would be that religion will only affect respondents' views if they participate in or are knowledgeable about their religion. In the United States, religiosity or commitment (in contrast to identity or belief) is often measured through attendance at religious services. Since during my data collection many religious services were shut down because of Covid, I had to replace this traditional measurement for my study. Instead, I asked a different question commonly used to measure strength of religiosity: "To what extent do you think of yourself as a religious person?" Possible answers were "Not religious at all" (1), "Slightly religious" (2), "Moderately religious" (3), and "Very religious" (4).[23]

I therefore removed the few assigned to the above Christian dichotomous variables who also claimed that they were "Not religious at all" or "Slightly religious" to a "Low-commitment Christian" dichotomous variable. In sum, I have dichotomous variables for Catholics (13.2 percent), conservative Protestants (21.9), liberal Protestants (5.2), identity-rejecting Protestants (7.0), low-commitment Christians (23.6), members of other religions (7.2), and the nonreligious (21.7) (the reference group in the models).

One question remained: Who are the identity-rejecting Protestants? Not much is known by researchers about this group.[24] Some scholars have observed that, in their theology and on social issues, this group lies somewhere between conservative and liberal Protestants.[25] However, in another paper I found that they are populist in their distrust of institutions. [26] This populism is consistent with a Protestantism that rejects theological traditions and doctrines, favoring personal experience over theories and doctrines, as in "seeker" conservative Protestant churches, which eschew labels and would just call themselves "Christian."[27]

Let us look at their use of these labels, which will tell us a lot about who they are. People who selected "Protestant" or "Just Christian" were sent to the more detailed question about Protestant identity. Of those who were eventually assigned to the "conservative Protestant" category, 76 percent first identified with "Protestant" and 24 percent with "Just Christian," whereas 92 percent of liberal Protestants identified with "Protestant" and only 8 percent with "Just Christian." This would be expected, since, because liberals are more committed to institutions, they would probably know the name of and identify with their specific denomination (e.g., Presbyterian Church (USA)).

But, of the identity-rejecting group, only 30 percent initially claimed "Protestant" and 38 percent "Just Christian," before eventually refusing to identify with any of the more specific Protestant identities. Thus, they definitely do not recognize their identity as "Protestant" and are more likely to identify with "Just Christian" than are conservative Protestants. Moreover, 32 percent chose "Something else" on the first question, which sent them to the write-in box. But those who selected "Protestant" or "Just Christian" were sent to a specific Protestant-identity question, and certain selections led me to categorize some (e.g., "evangelical," "fundamentalist," "born-again") as conservative Protestants, and some (e.g., "mainline Protestant," "liberal Protestant") as liberal Protestants. Respondents who selected either "Protestant" or "Just Christian" but did not identify with any of the specific Protestant identities were likewise sent to the write-in box. Thus, the only way respondents ended up in the identity-rejecting group was either by selecting "Protestant" or "Just Christian" in the first step and then rejecting the specific Protestant identities and leaving the write-in option blank, or by their response to the write-in box after having rejected the listed identities.

So who are these identity-rejecting Protestants? I used a number of other questions in an attempt to define them.[28] I asked respondents about their certitude about the existence of God. The choices were, in order, "I don't believe in God"; I don't know whether there is a God and don't believe there is any way to find out"; "I don't believe in a personal God, but I do believe in a Higher Power of some kind"; "I find myself believing in God some of the time, but not at others"; "While I have doubts, I feel that I do believe in God"; and "I know God really exists and I have no doubts." I found that the conservative Protestants and the identity-rejecting Protestants had the same average response to this question. I also asked the traditional GSS question on biblical authority, with its three response categories ranging from seeing the Bible as human-made to regarding it as the literal word of God. The conservative Protestants and the identity-rejecting Protestants had the same average response to this question. Thus, on these basic markers of conservative orthodoxy, the identity-rejecting Protestants resemble the conservative Protestants.

The largest differences are in demographics. The non-identity group is much less educated, much younger, much more female, and less affluent than the conservative Protestants.[29] The racial makeup of this group is even more distinct. While conservative and liberal Protestants have roughly the same racial makeup, this identity-rejecting group is 24 percent African American (compared to about 10 percent for both conservative and liberal Protestants), 13 percent Hispanic (compared to about 4 percent), and 10 percent Asian (compared to about 5 percent). This reminds us that these Protestant identity markers are more associated with white than with nonwhite Protestantism. Compared with the white conservative Protestants, the identity-rejecting Protestants are not as Republican.

The overall picture of identity-rejecting Protestants that I am drawing here is of people who are actively religious but are not engaged with traditional Protestant doctrine. The best evidence of this is that, despite having the same certitude about God and the same view of biblical authority as the conservative Protestants, they were much different in their responses to the one conservative-Protestant doctrinal question I asked, which is whether humans are those made in the image of God (the conservative-Protestant doctrinal interpretation of the first chapter in the first book of the Christian Bible). The identity-rejecting Protestants were much less likely than the conservative Protestants to exclusively believe in *Imago Dei* compared to other anthropologies.[30] Their theology is thus akin to what Christian Smith finds as the dominant religion of the young, which he calls "moralistic therapeutic deism."[31] Young Christians do believe—but that is about all they know. In this, their belief resembles the sort of stripped-down religion encountered in megachurches and seeker churches.

Such identity-rejecting Protestants know the public narrative about Christianity. They know that they are supposed to be biblical literalists and have certainty about the existence of God. But they are not aware of the history or doctrines of their faith; if they were, they would recognize that they are indeed "Protestants." If they are indeed members of megachurches or seeker churches, their clergy did not necessarily go to seminary to learn these doctrines. So these identity-rejecting Protestants can best be summarized as conservative Protestants without doctrine.

Statistical Analysis

For this survey I generally used the ordered logistic model (OLM). One assumption in an OLM is the *parallel regression* or *proportional odds assumption*.[32] This assumption is that the effect of a variable on the difference between the first category of the dependent variable and the rest of the categories is the same as the effect of the first and second combined on the remaining categories, and so on. If this assumption is violated, then the regular OLM coefficient becomes too much of a generalization and thus potentially misleading. It turns out that most OLM models violate this assumption for at least one variable. [33]

I diagnose the proportional odds assumption using the autofit function (p<.05) of the gologit2 program.[34] While it is possible to provide gologit2 results, there are four sets of results for a five-point dependent variable, compared to a single set for the regular OLM model. Since we do not need that level of detail, I only describe the gologit2 results in truncated form if the OLM is misleading.

A number of patterns can be seen in non-parallelism. One is that the effects change markedly between comparisons, but none of the comparisons are statistically significant. Another is that a general direction of an effect varies in strength as it goes up the scale. Since I consider neither of these to be substantively different from the OLM results, I do not report them. For the topic of this book, there is no theory so precise as to say that the effect of age on the difference between "Strongly disagree" and more agreeable answers will be 20 percent larger than the also significant effect of age on the difference between "Strongly disagree" and "Somewhat disagree" on the more agreeable answers.

I only mention non-parallelisms when they impact a specific hypothesis and are different from the more general OLM model. This typically occurs when an effect is non-significant at one end of the scale but grows or shrinks to significance at the other end of the scale. For example, a variable may not have an impact on people choosing "Strongly disagree" vs. "Somewhat disagree" and higher, but may have a very strong impact on choosing "Strongly disagree" through "Somewhat agree" vs. "Strongly agree." This is what the inventor of the program calls "asymmetrical effects," where, for example, a variable predicts disagreement but not agreement, or vice versa. [35] It is often the case that an effect is dichotomous, such as predicting "Strongly disagree" vs. more agreeing responses. Finally, since measurement of the concepts in this book are so difficult and noisy, I am going to report findings that are significant at p<.10 (two-tailed). I will distinguish these .10 effects from the p<.05 effects by calling them "weak effects." More technically, "weak" means "less confidence in the effect," but that phrase does not flow well in prose.

Manipulation Checks on Vignette Experiments

The vignette experiments are explained in the main text. Surveys that have an experimental design presume that the respondent notices the experimental stimulus. For example, in the vignette I use for NCs, some respondents read that the animal has brain tissue *like a human's* and some read that the brain tissue is *from a human*. Testing the hypothesis presumes that the respondent has internalized the idea that the brain tissue came from a human or is like a human's. This is one level more specific than generally paying attention to what you are reading with an attention check.

To confirm that the manipulation was internalized, studies use *manipulation checks* (MCs) to determine whether an experimental manipulation was effective. John V. Kane and Jason Barabas make useful distinctions between different types of manipulation checks that may be appropriate for different projects.[36] For each of the vignettes, I included a series of "factual" MCs in the screen following the dependent variable questions.[37] These are factual questions about an aspect of the vignette but worded in a different way. The respondent can only get it right by either guessing or by having internalized the factor in the dimension of the vignette. A manipulation is considered to have occurred when the check is statistically significantly related to the factor in question.[38] Unfortunately, it is not possible to conduct a manipulation check for every factor, both because too many questions would be required and because an appropriate question cannot always be formulated.

For the HBO vignette, the statement read, "We have a few more questions about the description you read. The description said that scientists have produced: a) One organoid; b) Ten thousand organoids; c) One hundred thousand organoids." Again, the correct answer depended upon the level of the factor, but 75.6 percent got this right. Another statement read, "The description said the debate is about whether an organoid would: a) Know it exists; b) Be aware of its surroundings." The correct answer was dependent upon the level, and 68.0 percent got this right.

For the NC vignette, a respondent who had seen one of the two brain-change levels then saw the statement, "These new animals would have: a) part of a brain from a human implanted in them; b) a genetically modified brain." Respondents who saw the immune system or movement factors were next presented with the statement, "I think it would be hard to notice that these animals are different than other [mice/monkeys] from watching them move: a) Agree; b) Disagree." While the respondents were presented with different statements depending on the factor level they read, 73.7 percent got this right.

Finally, to try to see if the "new species" level had been internalized, I asked, "How likely is it that you will eventually see one of these [humanized (mice/monkeys)/chimeras/Cebirs]? a) Definitely; b) Very probably; c) Probably; d) Possibly; e) Probably not; f) Definitely not." This is not a factual question but is subjective, and relies upon a reasonable prediction from what we know about how we experience the world. Presumably those who read that the chimera was a new species would have been more likely to believe that they will probably eventually see one because a species would be fairly common, compared to seeing that only a dozen were created.[39] All of the manipulation checks were statistically significant. Analysts are generally discouraged from dropping respondents who fail manipulation checks.[40]

Notes

Chapter 1

1. "1968 in Film," en.wikipedia.org/wiki/1968_in_film.
2. Robert Klitzman, "How Artistic Representation Can Inform Current Debates about Chimeras," *Journal of Medical Humanities* 42, no. 3 (September 1, 2021), 338.
3. Klitzman, 340.
4. en.wikipedia.org/wiki/The_Brain_That_Wouldn%27t_Die.
5. David J. Chalmers, "The Matrix as Metaphysics," in *Philosophers Explore the Matrix*, ed. Christopher Grau (New York: Oxford University Press, 2005), 132.
6. Jack-Yves Deschamps et al., "History of Xenotransplantation," *Xenotransplantation* 12, no. 2 (2005), 92, 95.
7. Deschamps et al., 99.
8. Deschamps et al., 97–99.
9. Kok Hooi Yap et al., "Aortic Valve Replacement: Is Porcine or Bovine Valve Better?" *Interactive CardioVascular and Thoracic Surgery* 16, no. 3 (March 1, 2013), 373.
10. Henry T. Greely et al., "Thinking about the Human Neuron Mouse," *American Journal of Bioethics* 7, no. 5 (May 10, 2007); Leonard D. Shultz, Fumihiko Ishikawa, and Dale L. Greiner, "Humanized Mice in Translational Biomedical Research," *Nature Reviews Immunology* 7, no. 2 (February 2007).
11. Tsutomo Sawai et al., "The Ethics of Cerebral Organoid Research: Being Conscious of Consciousness," *Stem Cell Reports* 13, no. 3 (September 2019), 443.
12. Giorgia Quadrato et al., "Cell Diversity and Network Dynamics in Photosensitive Human Brain Organoids," *Nature* 545, no. 7652 (May 2017), 53.
13. Elke Gabriel et al., "Human Brain Organoids Assemble Functionally Integrated Bilateral Optic Vesicles," *Cell Stem Cell* 28 (October 7, 2021), 1740.

14. Cleber A. Trujillo et al., "Complex Oscillatory Waves Emerging from Cortical Organoids Model Early Human Brain Network Development," *Cell Stem Cell* 25, no. 4 (October 2019), 6.

15. The first statement was from Alysson Muotri, the second from Gabriel Silva, both at the 2021 Sanford Stem Cell Symposium, October 14, 2021. Both are my colleagues on the faculty at the University of California, San Diego.

16. Bret Stetka, "Lab-Grown Mini Brains Can Now Mimic the Neural Activity of a Preterm Infant," *Scientific American* (August 29, 2019), http:www.sci entificamerican.com/article/lab-grown-mini-brains-can-now-mimic-the-neural-activity-of-a-preterm-infant/.

17. Jason Scott Robert and Françoise Baylis, "Crossing Species Boundaries," *American Journal of Bioethics* 3, no. 3 (August 2003), 1.

18. Greely et al., 29.

19. Nita A. Farahany et al., "The Ethics of Experimenting with Human Brain Tissue," *Nature* 556, no. 7702 (April 2018), 430; Xiaoning Han et al., "Forebrain Engraftment by Human Glial Progenitor Cells Enhances Synaptic Plasticity and Learning in Adult Mice," *Cell Stem Cell* 12, no. 3 (March 7, 2013), 342.

20. Insoo Hyun, J. C. Scharf-Deering, and Jeantine E. Lunshof, "Ethical Issues Related to Brain Organoid Research," *Brain Research* 1732 (April 2020), 4. The NASEM committee tasked with studying the ethics of HBOs and NCs (of which I was a member) advocated calling these entities "transplants." I think this confuses verb and noun, and I believe they fit into the NC category.

21. Omer Revah et al., "Maturation and Circuit Integration of Transplanted Human Cortical Organoids," *Nature* 610, no. 7931 (October 13, 2022), 319.

22. Carl Zimmer, "Human Brain Cells Grow in Rats, and Feel What the Rats Feel," *New York Times* (October 12, 2022).

23. Zimmer.

24. Zimmer.

25. Stuart Atkinson, "Neural Blastocyst Complementation: A New Means to Study Brain Development in Mouse," *Stem Cells Portal* (January 21, 2019), https://stemcellsportal.com/article-scans/neural-blastocyst-complementation-new-means-study-brain-development-mouse.

26. Alejandro De Los Angeles, Nam Pho, and D. Eugene Redmond, Jr., "Generating Human Organs via Interspecies Chimera Formation: Advances and Barriers," *Yale Journal of Biology and Medicine* 91, no. 3 (September 23, 2018), 334.

27. Amelia N. Chang et al., "Neural Blastocyst Complementation Enables Mouse Forebrain Organogenesis," *Nature* 563, no. 7729 (November 2018), 126.

28. Guoping Feng et al., "Opportunities and Limitations of Genetically Modified Nonhuman Primate Models for Neuroscience Research," *Proceedings of the National Academy of Sciences* (August 19, 2020), 24022–31.

29. Tao Tan et al., "Chimeric Contribution of Human Extended Pluripotent Stem Cells to Monkey Embryos Ex Vivo," *Cell* 184, no. 8 (April 15, 2021), 2020.

30. Nicole C. Walsh et al., "Humanized Mouse Models of Clinical Disease," *Annual Review of Pathology: Mechanisms of Disease* 12, no. 1 (2017), 187.

31. Anthony King, "The Search for Better Animal Models of Alzheimer's Disease," *Nature* 559, no. 7715 (July 25, 2018), S13; Zhen Liu et al., "Autism-like Behaviours and Germline Transmission in Transgenic Monkeys Overexpressing MeCP2," *Nature* 530, no. 7588 (February 2016), 98.

32. Antonio Regalado, "Chinese Scientists Have Put Human Brain Genes in Monkeys—and Yes, They May Be Smarter," *MIT Technology Review* (April 10, 2019).

33. Feng et al.

34. Regalado.

35. Andy Coghlan, "The Smart Mouse with the Half-Human Brain," *New Scientist* (December 1, 2014).

36. John H. Evans, *The History and Future of Bioethics: A Sociological View* (New York: Oxford University Press, 2012), xxx–xxxiii.

37. Evans, Chap. 2.

38. For example, the National Academies convened a committee to study the ethics of the HBOs and NCs. This document summarizes the entities that have produced "nonbinding guidance" about NC. This includes the National Research Council/Institute of Medicine report in 2010 and the International Society for Stem Cell Research report in 2016. It also talks about National Institutes of Health Policy, which itself refers to its ultimate arbiter, the Advisory Committee to the Director: Working Group for Human Embryonic Stem Cell Eligibility Review. See Committee on Ethical, Legal, and Regulatory Issues Associated with Neural Chimeras and Organoids et al., *The Emerging Field of Human Neural Organoids, Transplants, and Chimeras* (Washington, DC: National Academies Press, 2021), 116–18.

39. Evans, Chap. 2.

40. Dietram A. Scheufele et al., "What We Know about Effective Public Engagement on CRISPR and Beyond," *Proceedings of the National Academy of Sciences* 118, no. 22 (June 1, 2021).

41. Evans, Chap. 5.

42. In earlier work I identified a distinct definition I called the "socially conferred," which does not coherently exist in the academic debate. See Evans, *What Is a Human? What the Answers Mean for Human Rights* (Oxford University Press, 2016), Chap. 1. I do not include that definition in this work because, unlike the others, it does not define humans in relationship to animals.

43. Warren S. Brown, "Cognitive Contributions to Soul," in *Whatever Happened to the Soul? Scientific and Theological Portraits of Human Nature,* ed. Warren S. Brown, Nancey Murphy, and H. Newton Malony (Minneapolis, MN: Augsburg Press, 1998), 101–102.

44. Chris Degeling, Rob Irvine, and Ian Kerridge, "Faith-Based Perspectives on the Use of Chimeric Organisms for Medical Research," *Transgenic Research* 23, no. 2 (April 2014), 272, 275.

45. Charles Comosy and Susan Kopp, "The Use of Non-Human Animals in Biomedical Research: Can Moral Theology Fill the Gap?" *Journal of Moral Theology* 3, no. 2 (June 1, 2014), 65.

46. Karen Lebacqz, "Alien Dignity: The Legacy of Helmut Thielicke for Bioethics," in *Religion and Medical Ethics: Looking Back, Looking Forward,* ed. Allen Verhey (Grand Rapids, MI: Eerdmans, 1996), 46.

47. Hans Joas, *The Sacredness of the Person: A New Genealogy of Human Rights* (Washington, DC: Georgetown University Press, 2013), 143.

48. Evans, *What Is a Human?*, 53.

49. Peter Singer, "The Sanctity of Life," *Foreign Policy* (October 20, 2009), 40.

50. Michael Tooley, "Personhood," in *A Companion to Bioethics,* 2nd ed., ed. Helga Kuhse and Peter Singer (New York: Wiley-Blackwell, 2009), 133.

51. Singer, 40.

52. Robert and Baylis, 3.

53. John West, *Darwin Day in America* (Wilmington, DE: ISI Books, 2007), 3–4.

54. Andrea Lavazza and Marcello Massimini, "Cerebral Organoids: Ethical Issues and Consciousness Assessment," *Journal of Medical Ethics* 44, no. 9 (September 2018), 1; Tsutomu Sawai et al., "Ethics of Cerebral Organoid Research," 442.

55. Henry T. Greely, "Human/Nonhuman Chimeras: Assessing the Issues," in *Oxford Handbook of Animal Ethics,* ed. Tom L. Beauchamp and R. G. Frey (New York: Oxford University Press, 2011), 674–75.

56. Henry T. Greely, "Human Brain Surrogates Research: The Onrushing Ethical Dilemma," *American Journal of Bioethics* 21, no. 1 (2021), 34.

57. Greely et al., 35. "In Kafka's 'Metamorphosis,' Gregor Samsa was transformed into a cockroach; would these experiments, in any relevant way, transform a mouse into a man? Or, to be more precise, into a creature with some aspects of human consciousness or some distinctively human cognitive abilities?" The authors conclude that a mouse brain is too small to "have human attributes, including consciousness."

58. H. Isaac Chen et al., "Transplantation of Human Brain Organoids: Revisiting the Science and Ethics of Brain Chimeras," *Cell Stem Cell* 25, no. 4 (October 2019), 462, 465. A similar concern is associated with the "moral status framework" advocated by Robert Streiffer. In that framework, "what is distinctively problematic about chimera research is the possibility that the introduction of human material would enhance an animal's moral status to the level of a normal human adult without respecting the moral obligations entailed by that status." The problem is that "the animal might continue to be treated in ways typical of animal research subjects and which would be profoundly unethical given its new moral status." See Robert Streiffer, "Human/Non-Human Chimeras," in *The* Stanford Encyclopedia of Philosophy (Summer 2019 edition), ed. Edward N. Zalta, https://plato.stanford.edu/archives/sum2019/entries/chimeras/, 2019, 14.

59. Committee on Ethical, Legal, and Regulatory Issues Associated with Neural Chimeras and Organoids et al., 4–5.

60. Carl Zimmer, "Organoids Are Not Brains. How Are They Making Brain Waves?" *New York Times* (August 29, 2019).

61. Sharon Begley, "Tiny Human Brain Organoids Implanted into Rodents, Triggering Ethical Concerns," STAT (November 6, 2017).

62. Ian Sample, "Growing Brains in Labs: Why It's Time for an Ethical Debate," *Guardian* (April 25, 2018).

63. Greely, "Human Brain Surrogates Research," 34.

64. There are many overlapping and reinforcing reasons; I will gesture at a few, while recognizing that a full examination would require a book-length text of its own.

    First, as I mentioned above, seeing life-forms on a continuum in which there are descendants of the same original life-form is the domain assumption in biology. While the continuum is based on evolution, it coincides with valued capacities. The worm is at the top of neither the evolutionary tree nor the continuum of valued capacities. Any bioethical debate that is going to have influence on an elite level needs to make sense to scientists.

Second, the philosophical anthropology is nearly consensual in the discipline of philosophy, and given that philosophy is the training ground for bioethicists, bioethicists will be trained to see this philosophical anthropology as taken for granted.

Third, the focus on capacities like consciousness allows scientists' expertise to be relevant to the debate. In principle, scientists can use science to determine whether an entity has the requisite level of consciousness, but scientists are not qualified to weigh in on the approaches to evaluating humanhood I will enumerate in subsequent pages.

Fourth, levels of consciousness among animals is a problem scientists are familiar with. Different assumed levels of consciousness, or more specific capacities that are thought to flow from consciousness, like feeling pain, determine which animals you can experiment on and how you are to treat these animals. For example, if you are experimenting on invertebrates like worms you do not even need to take your research to an oversight committee, but the oversight committee will be more stringent the more humanlike the animal is.

Fifth, this focus on consciousness fits well with the individualism not only assumed by bioethics and science but essentially required in law. That is, the question is whether we are harming this one individual mouse. The answer is yes if we upgrade its consciousness and still experiment upon it. This individualizing ethical problems is in contrast to other classes of ethical objection I will consider below that are not addressed by bioethics, since they are ultimately not about individuals but about society or culture.

Chapter 2

1. Evans, *History and Future of Bioethics*, Chap. 5.
2. Moreover, academics see anthropologies as mutually exclusive, whereas ordinary people typically adhere to multiple anthropologies, simply tending toward one or the other. Unlike academics, the public is not rewarded for being logically consistent. See Evans, *What Is a Human?*
3. Evans, 18–19.
4. Evans, 53.
5. Evans, 137.
6. Evans, 102–3.
7. Evans, 93.

8. Anthropology and this human exceptionalism are related. The more respondents agreed with the biological or philosophical anthropology, the *less* likely they were to agree that humans are special compared to animals. The more respondents agreed with the theological anthropology, the more likely they were to agree that a human is special compared to animals. (See Evans, 60.) This is consistent with the academic debate in which those relying on the philosophical and biological anthropology see humans and animals on the same continuum. Chimpanzees, which are often considered by academics to have personhood, have lower status than humans. One question I asked in the in-depth interviews in the earlier project was, "Should we kill a chimpanzee if it were necessary to create a medical treatment to save human lives?" It is notable that only one of 92 interviewees said no. Nearly everyone else wanted to avoid harming the chimpanzee if possible, but in the end, if forced to choose, considered humans to be more valuable than chimpanzees.

9. Dietmar Hübner, "Human-Animal Chimeras and Hybrids: An Ethical Paradox behind Moral Confusion?," *Journal of Medicine and Philosophy* 43, no. 2 (March 13, 2018), 195. He continues, "It is thus a form of partiality, analogous to racism or sexism, differentiating between things not with respect to certain general properties with definite moral weight, but only with regard to their individual affiliation to some arbitrary group—that is, without providing relevant reasons for the distinction." Of course, this debate is over what "relevant reasons" consist of.

10. While I believe my previous study is the only formal empirical study on the public's definition of a human, there are more general indicators of the public's views that are consistent with my analysis, at least in the distinction between humans and animals. General reflection on an analyst's personal experience often reaches this conclusion. For example, Robert and Baylis write that "the putative fixity of putative species boundaries remains firmly lodged in popular consciousness and informs the view that there is an obligation to protect and preserve the integrity of human beings and *the* human genome" ("Crossing Species Boundaries," 10).

   We would also expect law to at least be somewhat based on public opinion, and indeed the legal definition of a human mirrors the view I find among the public. Bartha Maria Knoppers and Henry Greely describe the classical legal definitions of a human, obviously formulated before the existence of modern biology: "The law starts with the position that any living organism born from a person is a natural person from at least the time of birth until the time of death." Moreover, they continue, an agency of the

United Nations "maintained that it is the human genome that 'underlies the fundamental unity of all members of the human family as well as the recognition of their inherent dignity and diversity.'" See Knoppers and Greely, "Biotechnologies Nibbling at the Legal 'Human,'" *Science* 366, no. 6472 (December 20, 2019), 1455.

11. John D. Loike, "Opinion: Should Human-Animal Chimeras Be Granted 'Personhood'?," *The Scientist* (May 23, 2018), 3. Loike points out that both the Bible and the Talmud assert that being born of a human is the intrinsic definition of a human being. Since Adam was not born, God had to declare Adam a human being; the Golem, not being born from a human, was not considered a human.

12. Robert and Baylis. Those familiar with these debates will note that Robert and Baylis self-identify as bioethicists. These two are not, however, mainstream bioethicists, and are part of a minority that use assumptions differing from those of the average bioethicist I discussed previously. For example, Baylis describes herself as a critic of mainstream bioethics. As her website states, "Her work . . . aims to move the limits of mainstream bioethics and develop more effective ways to understand and tackle public policy challenges in Canada and abroad," 1. www.dal.ca/sites/noveltec hethics/our-people/francoise-baylis.html

13. Robert and Baylis, 2, 8.

14. Robert and Baylis, 10. I suspect that the vaguer complaint about these technologies being "unnatural" is a stand-in for not respecting the (natural) divide between humans and other life-forms. See Phillip Karpowicz, Cynthia B. Cohen, and Derek van der Kooy, "It Is Ethical to Transplant Human Stem Cells into Nonhuman Embryos," *Nature Medicine* 10, no. 4 (April 2004), 332. I lack specific data for this conclusion, but simply note that essentially no one says it is "unnatural" to modify nature through medicine or building homes, no one says it is "unnatural" to crossbreed plants, and few would say it is "unnatural" to crossbreed different non-human animals. Rather, the alterable aspect of "nature" that everyone cares about is the human.

15. Committee on Ethical, Legal, and Regulatory Issues Associated with Neural Chimeras and Organoids et al., *Emerging Field*, 52–54.

16. Maartje Schermer, "The Mind and the Machine: On the Conceptual and Moral Implications of Brain-Machine Interaction," *NanoEthics* 3, no. 3 (December 2009), 218; Tsjalling Swierstra, Rinie van Est, and Marianne Boenink, "Taking Care of the Symbolic Order: How Converging Technologies Challenge Our Concepts," *NanoEthics* 3, no. 3 (December 2009), 270.

17. Mary Douglas, *Purity and Danger: An Analysis of the Concepts of Pollution and Taboo* (1966; repr.: London: Routledge, 1984); Martijntje Smits, "Taming Monsters: The Cultural Domestication of New Technology," *Technology in Society* 28, no. 4 (November 2006), 493.

18. Douglas, Chap. 3.

19. Jeffrey Stout, "Moral Abominations," *Soundings* 66, no. 1 (1983), 9.

20. Matthew Engelke, *How to Think Like an Anthropologist* (Princeton, NJ: Princeton University Press, 2019), 26–28.

21. Stout, 7, 8.

22. Helen De Cruz and Johan De Smedt, "The Role of Intuitive Ontologies in Scientific Understanding—the Case of Human Evolution," *Biology & Philosophy* 22, no. 3 (June 2007), 352–53.

23. Timothy M. Renick, "A Cabbit in Sheep's Clothing: Exploring the Sources of Our Moral Disquiet about Cloning," *Annual of the Society of Christian Ethics* 18 (1998), 265.

24. Gill Haddow, "Animal, Mechanical, and Me," *Oxford Handbook of the Sociology of Body and Embodiment,* ed. Natalie Boero and Katherine Mason (New York: Oxford University Press, 2020), 174–75. See more generally Haddow, *Embodiment and Everyday Cyborgs: Technologies That Alter Subjectivity* (Manchester, Eng.: Manchester University Press, 2021).

25. Duschinsky, 1.

26. Smits, 495.

27. Robbie Duschinsky, "Introduction," *Purity and Danger Now: New Perspectives*, ed. Robbie Duschinsky, Simone Schnall, & Daniel H. Weiss (New York: Routledge, 2016), 7.

28. Smits, 501.

29. Distinctions are necessary because "separating entities from their surroundings is what allows us to perceive them in the first place," and categorization is a central aspect of learning for the young. See Eviatar Zerubavel, *The Fine Line: Making Distinctions in Everyday Life* (New York: Free Press, 1991), 1; Renick, 264. For example, children learn that a bird is somehow in a different category than a cow.

30. Zerubavel, *The Fine Line*, 51–55. Particular institutions within a society also teach the importance of strong distinctions in general, although not necessarily the foundational human-animal one that concerns us here. For example, law is a "hotbed of rigidity" in that it admonishes lawyers for being "wishy-washy" or "fuzzy." Science "dreads anomalies" and tries to cram all observations into categorical paradigms until . . . the paradigm explodes and is replaced by a new and similarly rigid one. See Zerubavel, *The Fine Line*, 59.

31. Zerubavel, *Social Mindscapes: An Invitation to Cognitive Sociology* (Cambridge, MA: Harvard University Press, 1997), 57–58.

32. Zerubavel, *The Fine Line*, 106.

33. Michael Morrison and Stevienna de Saille, "CRISPR in Context: Towards a Socially Responsible Debate on Embryo Editing," *Palgrave Communications* 5, no. 1 (December 2019), 5.

34. Amy Hinterberger, "Regulating Estrangement: Human-Animal Chimeras in Postgenomic Biology," *Science, Technology, & Human Values* 45, no. 6 (November 2020), 1070.

35. Peter Morriss, "Blurred Boundaries," *Inquiry* 40, no. 3 (September 1997), 275. Similarly, Jeffrey Stout writes that "the more severe the anomaly or ambiguity at hand, the more paradoxical and mysterious our speech is apt to become. This helps explain not only the potentially emotional character of our revulsion, given the uneasiness that can be produced whenever coherent interpretation of experience is impossible, but also the suspicion that the very notion of abomination is somehow beyond comprehension." See Stout, "Moral Abominations," 16.

36. Mike R. King, Maja I. Whitaker, and D. Gareth Jones, "I See Dead People: Insights from the Humanities into the Nature of Plastinated Cadavers," *Journal of Medical Humanities* 35, no. 4 (December 2014), 362.

37. Greely, "Human/Nonhuman Chimeras," 682.

38. Morriss, 275.

39. Greely, 687.

40. Academy of Medical Sciences, *Animals Containing Human Material* (London: Academy of Medical Sciences, 2011), 3.

41. There is a research tradition about people's sense of unease in the appearance of robots. The basic thesis is that people have an increasingly favorable reaction to increasingly human-appearing robots, but that robots that are very humanlike yet not fully human-appearing are the most unsettling. There is a zone of not-quite-human-looking that is the most unsettling. People are not unsettled by a robot that is a metal cube or by one that is indistinguishable from a human. See Jari Katsyri et al., "A Review of Empirical Evidence on Different Uncanny Valley Hypotheses: Support for Perceptual Mismatch as One Road to the Valley of Eeriness," *Frontiers in Psychology* 6 (April 10, 2015).

Recent research shows that it is aspects of the face that trigger the unsettled response. Indeed, in the words of one review, "human faces are probably the most important visual stimuli in our social environment" and there appear to be brain functions for "special processing" of faces. This makes sense in terms of human evolution, in which the ability to distinguish different faces would have given our ancestors reproductive advantage. More importantly, this processing is specific to your own species. See Kun Guo, David Tunnicliffe, and Hettie Roebuck, "Human Spontaneous Gaze Patterns in Viewing of Faces of Different Species," *Perception* 39, no. 1 (April 1, 2010), 533. Since the human face is probably the nexus of appearance-related identity, a human face would produce the most extreme sense of violation of the human-animal distinction.

42. Evans, *What Is a Human?*, Chap. 5.

43. Sarah Parry, "Interspecies Entities and the Politics of Nature," *Sociological Review* 58, no. 1, suppl. (May 2010), 121.

44. Morriss, 269.

45. Morriss, 269.

46. Greely et al., 34.

47. Sara Worley, "Materialism versus Dualism," in *Encyclopedia of Clinical Psychology*, ed. Robin L. Cautin and Scott O. Lilienfeld (New York: John Wiley, 2015), 1.

48. Quoted in Brian Dolan, "Soul Searching: A Brief History of the Mind/Body Debate in the Neurosciences," *Neurosurgical Focus* 23, no. 1 (July 2007), 5.

49. Alvin Plantinga, "Against Materialism," *Faith and Philosophy* 23, no. 1 (February 1, 2006), 3.

50. Plantinga, 3.

51. Evans, 134, 138.

52. Evans, 145.

53. Evans, 148, 149.

54. Rebecca Skloot, *The Immortal Life of Henrietta Lacks* (New York: Broadway Books, 2017).

55. Renee C. Fox and Judith P. Swazey, *Spare Parts: Organ Replacement in American Society* (New York: Oxford University Press, 1992), Chap. 2.

56. Catherine Waldby et al., "Blood and Bioidentity: Ideas about Self, Boundaries and Risk among Blood Donors and People Living with Hepatitis C," *Social Science & Medicine* 59, no. 7 (October 1, 2004), 1463.

57. Skloot, 196, 262–63, 266.

58. Skloot, 295–96.

59. Waldby et al., 1467.

60. Waldby et al., 1466.

61. Waldby et al., 1466.

62. Sandra S.-J. Lee et al., "'I Don't Want to Be Henrietta Lacks': Diverse Patient Perspectives on Donating Biospecimens for Precision Medicine Research," *Genetics in Medicine* 21, no. 1 (January 2019), 110.
63. Fox and Swazey, 36.
64. Waldby et al., 1464.
65. Haddow, "Animal, Mechanical, and Me," 165.
66. Charlotte Ikels, "The Anthropology of Organ Transplantation," *Annual Review of Anthropology* 42, no. 1 (2013), 95.
67. Haddow, Embodiment and everyday cyborgs, Chap. 1.
68. Waldby et al., 1466–67.
69. Fox and Swazey, 38.

Chapter 3

1. For example, of the studies of the United States, one study is not of the public but rather of eight expert narratives on the topic. See Chris Degeling, Rob Irvine, and Ian Kerridge, "Faith-Based Perspectives on the Use of Chimeric Organisms for Medical Research," *Transgenic Research* 23, no. 2 (April 2014). An unrepresentative 2020 study of views of human-animal chimeric embryo research found that 59 percent of the respondents accept injecting human-induced pluripotent stem cells into genetically modified swine embryos and transplanting the resulting human tissues into a human. But it is unclear what subset of Americans this study represents. See Andrew T. Crane et al., "The American Public Is Ready to Accept Human-Animal Chimera Research," *Stem Cell Reports* 15, no. 4 (October 13, 2020), 808; Isabel Bolo, Ben Curran Wills, and Karen J. Maschke, "Public Attitudes toward Human-Animal Chimera Research May Be More Complicated Than They Appear," *Stem Cell Reports* 16, no. 2 (February 2021).

2. Acceptance of the use of animals in research increases with the respondent's age, male gender, rural residence, not owning a pet, being Christian, and particularly being Protestant. Unsurprisingly, people are more opposed to experimenting on animals that are kept as pets, animals that are more attractive to humans, and animals with "higher" mental abilities. People are more opposed to research that will do more harm to the animal, or that does not have the goal of advancing human health, and for which other means than experimenting on animals can be substituted. See Elisabeth H. Ormandy and Catherine A. Schuppli, "Public Attitudes toward Animal

Research: A Review," *Animals* 4, no. 3 (September 2014). In another study, "people tended to accept (or at least to tolerate) the suffering of the animals concerned when there existed a genuine and authentic human need, typically expressed in the need to cure life-threatening diseases." See Phil Macnaghten, "Animals in Their Nature: A Case Study on Public Attitudes to Animals, Genetic Modification and 'Nature,'" *Sociology* 38, no. 3 (2004), 540.

One step closer to the topic of this book is the large empirical literature on nonneural chimeras, such as pigs with organs that could be transplanted to humans. This could be directly relevant to people's views of breaking the boundary between humans and animals, albeit with parts not very associated with humanness. Upon closer examination, the vast majority of these studies survey people in specific occupations (e.g., Spanish nursing students) or make no effort at making generalizable claims about the US population.

In one qualitative British study, respondents were asked about transferring a human gene into an animal for a range of purposes, including xenotransplantation. "The dominant response in all groups was negative." See Macnaghten, "Animals in Their Nature," 544. One meta-analysis of these studies finds that about half of respondents accept xenotransplantation. See J. Hagelin, "Public Opinion Surveys about Xenotransplantation," *Xenotransplantation* 11, no. 6 (2004).

There is one representative survey of the US population. The title of the Pew survey report says it all: "Most Americans accept genetic engineering of animals that benefits human health, but many oppose other uses." When asked to react to producing "animals to grow organs/tissues for humans needing a transplant," 41 percent say that this is "taking technology too far," and 57 percent say it is an "appropriate use of technology." Men compared to women (65–49 percent), those with a higher level of science knowledge compared to a lower (72–47 percent), and those with low religious commitment compared to high (68–48 percent) are more likely to say it is appropriate. Pew finds that 52 percent of Americans oppose the use of animals in research, and these people, not surprisingly, are more opposed to chimeric animals (47–69 percent). See Cary Funk and Meg Hefferon, "Most Americans Accept Genetic Engineering of Animals That Benefits Human Health, but Many Oppose Other Uses," (August 16, 2018), https://www.pewresearch.org/science/2018/08/16/most-americans-accept-gene tic-engineering-of-animals-that-benefits-human-health-but-many-opp ose-other-uses/.

3. The interview data comes from the process of designing the survey. After drafting the survey I used cognitive testing to ensure that respondents would understand the questions in the way I intended. See Debbie Collins, *Cognitive Interviewing Practice* (Los Angeles: SAGE Publications, 2014); Camilla Priede and Stephen Farrall, "Comparing Results from Different Styles of Cognitive Interviewing: 'Verbal Probing' vs. 'Thinking Aloud,'" *International Journal of Social Research Methodology* 14, no. 4 (July 2011). Cognitive testing, a specific type of in-depth interviewing focused on the words of each survey question, is particularly important when asking questions about novel phenomena (as is the case here). One of the dominant techniques is called "think aloud," in which "the respondent is encouraged to vocalise their thought processes as they answer survey questions" See Priede and Farrall, 272. Traditional cognitive interviewing interprets the respondent's thought process and uses it to determine whether the respondent understood the question as intended.

   I took advantage of the fact that the core of my survey consists of vignette questions (of which more below), and the primary response question is about agreeing or disagreeing with the development of HBO or NCs. Asking a respondent to "think aloud" about a technology is essentially how I would have conducted an in-depth interview study. So, in addition to ensuring that the questions were understood as intended, I have in-depth interview data from 35 respondents who were diverse in gender, educational level, and religious tradition (although certainly not representative). I inductively coded key parts of those in-depth interviews using the interview coding program NVivo. Given that the survey is the touchstone for both methods, I put most of the interpretive weight on the survey.

4. John H. Evans and Justin Feng, "Conservative Protestantism and Skepticism of Scientists Studying Climate Change," *Climatic Change* 121, no. 4 (December 1, 2013), 607.

5. Diana C. Mutz, *Population-Based Survey Experiments* (Princeton, NJ: Princeton University Press), 9.

6. Mutz, 54, 56.

7. Complicated vignettes are a challenge to explain in a publication. This one had to account for the factors I really wanted to examine as well as small grammatical issues that had to be accommodated, such as referring to a single NC or HBO as "it" and more than one as "they."

8. Knowledgeable readers will note that some of the vignette combinations generated through random assignment are probably always going to be

scientifically impossible. For example, it is extremely unlikely that an HBO could be grown to 5 inches in size. As the saying in sociology goes, the truth of the matter does not matter. In attempting to understand people's reactions to HBO, it was sometimes necessary to create somewhat fictional scenarios so as to establish a great-enough difference so that measurement is possible. Nevertheless, I've resisted extreme fictions, like the possibility of HBOs controlling robots.

9. The wording combines these two factors into four statements, each randomly assigned to 25 percent of respondents. The actual factors shown were:

"[One was grown by reprogramming stem cells from the skin of the scientist doing this research. It is /One was manufactured by reprogramming the stem cells from the skin cells from the skin of the scientist doing this research. It is /Ten thousand were grown by reprogramming stem cells from the skin of the scientist doing this research. They are /Ten thousand were manufactured by reprogramming stem cells from the skin of the scientist doing this research. They are]."

10. Tsutomu Sawai et al., Using other terms from this debate, I am making a distinction between what Robert Van Gulick calls "self-consciousness," which is "a more demanding sense" of consciousness, and "sentience," which is "sensing and responding to the world," 443. Van Gulick, "Consciousness," in *The Stanford Encyclopedia of Philosophy*, ed. Edward N. Zalta and Uri Nodelman, Spring 2018 (Stanford, CA: Metaphysics Research Lab, Philosophy Department, Stanford University, 2018), 4. Humans and many more advanced animals, including primates and dolphins, are thought to have self-consciousness, whereas fish and lower animals have sentience, as I have defined it here.

11. Van Gulick.

12. The wording combines these two factors—what consciousness is and what the level of consciousness would be, resulting in eight distinct combinations that the respondent could be assigned to see. The first four describe consciousness as self-awareness, the later four describe it as awareness of surroundings:

"[self-awareness—knowing that it exists. The conclusion so far is that an organoid will not have self-awareness/self-awareness—knowing that it exists. The conclusion so far is that an organoid could have as much self-awareness as an insect does/self-awareness—knowing that it exists. The conclusion so far is that an organoid could have as much self-awareness as a pig does/self-awareness—knowing that it exists. The conclusion so far is that an organoid could have as much self-awareness <u>as a human fetus</u> does/[<u>awareness of its surroundings—know that things are around it</u>. The conclusion so far is that an organoid <u>will not be aware</u> of its surroundings/<u>awareness of its surroundings—know that things are around it</u>. The conclusion so far is that an organoid could have as much awareness of its surroundings <u>as an insect</u> does/<u>awareness of its surroundings—know that things are around it</u>. The conclusion so far is that an organoid could have as much awareness of its surroundings <u>as a pig</u> does/<u>awareness of its surroundings—know that things are around it</u>. The conclusion so far is that an organoid could have as much awareness of its surroundings <u>as a human fetus</u> does]."

13. All names are pseudonyms. Respondents whose name was typically associated with one gender (e.g., Mitch) were assigned pseudonyms typically associated with the same gender (e.g., Joseph).

14. No one in my in-depth interviews selected "Strongly disagree." That makes sense, because if the frequencies in the survey applied to my interviews, I would expect 1.5 respondents (.043 × 35 interviews) to select this response.

15. I recognize that one of the possible levels in one of the later factors precludes germline modification, but likely no respondents will be aware of this.

16. Christopher J. Preston, *The Synthetic Age: Outdesigning Evolution, Resurrecting Species, and Reengineering Our World* (Cambridge, MA: MIT Press, 2018).

17. Academy of Medical Sciences, *Animals Containing Human Material* (London, 2011), 6.

18. Feng, Jensen, et al.

19. For a theory of how these deep assumptions come into being, see Peter L. Berger and Thomas Luckmann, *The Social Construction of Reality: A Treatise in the Sociology of Knowledge* (New York: Anchor, 1967).

Chapter 4

1. Tables are available on my website, johnhevans.ucsd.edu. See Table 2, Column 1. For vignette analysis, complicated statistics are not required, so I present the differences in mean response to the general approval of the research question as modeled by a simple ordered logistic regression where the levels are dummy variables. "No consciousness" is the reference group.

Experts in social statistics should note that I use two related regression methods—the ordered logistic and the generalized ordered logistic. See Richard Williams, "Understanding and Interpreting Generalized Ordered Logit Models," *Journal of Mathematical Sociology* 40, no. 1 (January 2, 2016). Some characteristics of the respondents influence the difference between all of their choices (e.g., between "Strongly disagree" and "Somewhat disagree," or between "Somewhat disagree" and "Neither agree nor disagree"). Sometimes the characteristic only influences one dichotomous divide in response (e.g., between "Strongly disagree" and all other responses). I will label the latter a *binary effect*. The exact difference will be reported in an endnote. What I technically mean by this can be found in the statistical-analysis section of the Methodological Appendix. This distinction rarely results in a substantively different interpretation, but when it does I will report it in an endnote. Basically, if a binary effect divides most of the responses, that is a substantive finding. If the binary effect only affects a difference between the bulk of the responses and an extreme and relatively rare response at one end of the scale, the most accurate approach is to describe the regular result, which would be non-significance.

2. Seeing "human fetus" compared to "no consciousness" is associated with a binary version of less approval of HBOs. However, seeing "human fetus" is only associated with selecting "Strongly disagree" compared to all the more agreeing responses combined. Since very few people selected "Strongly disagree" (4 percent of respondents), this level predicts an extremely rare extremist response that is not representative of the public's views more generally. Therefore, the primary finding is "No effect." I suspect that using "human fetus" has sent a very small percentage of respondents into the abortion debate, and the very strong anti-abortion advocates did not simply weigh the word *fetus* in their decision but rather responded with an immediate and categorical "Strongly opposed."

3. There is a weak effect ($p = .06$) of the factor describing two types of awareness. An important question in debates about consciousness is whether consciousness refers to being aware that you exist or being aware of your surroundings, with self-awareness being the more human quality and thus more advanced version. I am making a distinction between what is called "self-consciousness," which is "a more demanding sense" of consciousness, and "sentience," which is "sensing and responding to the world." See Van Gulick, "Consciousness."

One of the factors in the vignette was designed to get at this distinction. If people's view of HBOs is structured by the level of consciousness the

HBO could obtain, then the more advanced self-awareness should generate disapproval. However, those who saw HBOs described as potentially becoming aware of their surroundings were more opposed than those who saw HBOs as potentially aware of themselves, which is the opposite of what the consciousness theory of support and opposition to HBOs would predict. This is a small difference, with "self-aware" predicting 72.9 percent "Agree" or "Strongly agree" and "aware of surroundings" 66.1 percent. This finding does not support the consciousness theory, but I am unsure of an interpretation for the negative finding.

4. For descriptive statistics for all of the standard survey questions, see Table 1 in the online Tables.

5. More technically for those familiar with social statistics, I use predicted values generated with the MTable command in Stata. For the exemplar analysis, I need to pick a hypothetical person and see how the person's answer would change if his or her response on the question of interest were to be different. I use the average response to the continuous response questions (e.g., age) to set the values I am not interested in, but pick certain categories for which an average is not logical (e.g., gender). For the categorical identities I set my analysis for the rest of this book at female, white, conservative Protestant, and Republican. See J. Scott Long and Jeremy Freese, *Regression Models for Categorical Dependent Variables Using Stata*, 3rd ed. (College Station, TX: Stata Press, 2014), 155.

6. See Table 3, Column 3. The difference between the most and least agreeing with the monkey question is 10 percent in agreeing or strongly agreeing with creating NCs. The top and bottom of the "humans and animals are similar" question have a 15-percent difference in agreeing or strongly agreeing with the creation of NCs.

7. See Table 4, Column 1.

8. See Table 4, Column 1. The "brain tissue like a human" level of this factor has a binary effect. It is associated with strongly disagreeing with creating NCs compared to disagreeing or more agreeing responses, as well as with strongly disagreeing or disagreeing and more agreeing responses. That is, the genetically modified brain invokes really strong disapproval and the distinction between disapproval and approval in general.

9. The intention was also to compare those who saw that tissue was being implanted from a human and those who saw that the animal brain was to be genetically modified to be like a human. While this is a very subtle difference, the former should induce more of a sense of breaching boundaries (because it is a combination of an actual existing human) than the latter (which is just making things similar). Comparing the two brain

modifications in the regression model by making the implant version the reference group does not produce a statistically significant difference (not shown).

10. While 43 percent of respondents who saw "change immune system" somewhat or strongly agreed with NC research, 34 percent of those who saw "move like a human" somewhat or strongly agreed. This is a predicted value with the other values set at mouse, new species, and name as "humanized" [monkey/mouse].

11. See Table 4, Column 1. The effect has a similar magnitude as does seeing the NC as now having human movement.

12. Committee on Ethical, Legal, and Regulatory Issues Associated with Neural Chimeras and Organoids et al., 142.

13. Seeing "monkey" instead of "mouse" produced a very weak increase in the likelihood of saying that there is not much of a difference between the NC and a human. See Table 4, Column 2. It is not that people did not notice that the animal had been changed, as this factor had a strong impact on general approval of the research in an earlier analysis. My interpretation is that, when we are talking about the human-animal distinction, the animal in question is not very important given the public's overwhelming desire to draw the distinction. (Only 18 percent of respondents agreed that there was not much of a difference between the NC and a human.)

14. This is a strong association. A cross-tabulation shows that 25 percent of those most strongly disagreeing that HBOs can be destroyed are somewhat or strongly agreeing with HBO research, whereas 83 percent of those who most agree that HBOs can be destroyed are equally agreeing with the research.

15. Table 2, Column 2.

16. See Table 4, Column 3.

17. The effect of seeing "Cebir" is in the same direction but not statistically significant. The chimera difference is binary, between "Strongly disagree," "Somewhat disagree," and "Neutral" compared to "Somewhat agree" and "Strongly agree." That is, it leads to people agreeing or disagreeing in general, but not to fine-grained distinctions within agreeing or disagreeing.

18. This is a binary effect with the difference between "Strongly disagree" and more agreeing categories, as well as between 'Strongly disagree" and "Somewhat disagree" compared to more agreeing categories.

19. See Table 4, Column 4. The difference between the reference group (immune system) and genetically modified brain is actually binary and is associated with selecting "Strongly agree" compared to "Somewhat agree," "Neutral," "Somewhat disagree," and "Strongly disagree." In other words,

there is no effect of seeing genetically modified brain except for those who are among the 7 percent of the sample who strongly agree that the entity be treated like a human. This effect in the tail of a distribution does not distract from the main finding of "No effect."

20. The "chimera" effect is in the same direction but not statistically significant.

21. The responses to this question and the other one in this section are not correlated with the amount of time it took for the respondent to complete the survey. Moreover, as we will see below, the questions are associated with a range of other variables in organized patterns, suggesting that these are thoughtful responses.

22. M. Susan Lindee and Dorothy Nelkin, *The DNA Mystique: The Gene as a Cultural Icon* (New York: W. H. Freeman, 1995), Chap. 8.

23. The model included the measure of general approval of HBO research to try to detangle extension from support. See Table 5, Column 1. The percentage agreeing is derived from an expected value analysis.

24. While this seems an extreme difference, the size of the difference is not due to my expected value calculation. These two questions are correlated at .558; the ordered logistic coefficient is .992 and has a z-value of 13.2.

25. The two ephemeral-connection questions were combined into an index with an Alpha of .65. The measures used in this analysis will be explained in the next chapter. See Table 7.

26. See Table 5, Column 3. The effect of thinking HBOs would have thoughts is binary, with the difference between the 9 percent of the sample who strongly agree with thoughts and all other responses. This is close to my cutoff for not being an actual effect in that it is modeling a tail of the distribution, but not an extreme tail.

27. See Table 5, Column 4.

Chapter 5

1. See Table 1 for the descriptive statistics of these variables and the Appendix for the religion variables that were measured.

2. John H. Evans and Justin Feng, "Conservative Protestantism and Skepticism of Scientists Studying Climate Change," *Climatic Change* 121, no. 4 (December 1, 2013); Gordon Gauchat, "The Political Context of Science in the United States: Public Acceptance of Evidence-Based Policy and Science Funding," *Social Forces* 94, no. 2 (December 2015); Gordon Gauchat, Timothy O'Brien, and Oriol Mirosa, "The Legitimacy of Environmental Scientists in the Public Sphere," *Climatic Change*

143, nos. 3–4 (August 2017); David Johnson, Christopher Scheitle, and Elaine Ecklund, "Individual Religiosity and Orientation towards Science: Reformulating Relationships," *Sociological Science* 2 (2015).

3. See Table 3, Column 3, and Table 6, Column 3.

4. John H. Evans and Eszter Hargittai, "Who Doesn't Trust Fauci? The Public's Belief in the Expertise and Shared Values of Scientists in the COVID-19 Pandemic," *Socius* 6 (January 1, 2020), 9.

5. The education effect is binary, with increased education disproportionally associated with selecting "Somewhat disagree" or "Strongly disagree" compared to the more agreeing responses.

6. The HBO and NC results are both binary. See Table 3, Column 3, and Table 6, Column 3.

7. Claire E. Altman, "Age, Period, and Cohort Effects," in *Encyclopedia of Migration*, ed. Frank D. Bean and Susan K. Brown (Dordrecht: Springer Netherlands, 2015).

8. Christopher J. Preston, *The Synthetic Age: Outdesigning Evolution, Resurrecting Species, and Reengineering Our World* (Cambridge, MA: MIT Press, 2018), 65.

9. Gregory E. Kaebnick, *Humans in Nature: The World as We Find It and the World as We Create It* (New York: Oxford University Press, 2013), 3.

10. Committee on Gene Drive Research in Non-Human Organisms et al., *Gene Drives on the Horizon: Advancing Science, Navigating Uncertainty, and Aligning Research with Public Values* (Washington, DC: National Academies Press, 2016), 63.

11. Committee on Gene Drive Research in Non-Human Organisms et al., 73.

12. Christopher J. Preston, xviii–xix. Control over nature for human use—the conservation perspective—is the driving motivation behind science and technology, including the development of HBOs and NCs. Robinson writes that "control over nature as the principal legitimating goal of science and reason was a chief impetus for the turn toward the empiricism that informed the work of Francis Bacon and others, and has been at the heart of scientific method, and, in some form, at the heart of the epistemological commitment of most scientists since." See Daniel N. Robinson, "Introduction," in *Scientism: The New Orthodoxy*, ed. Richard N. Williams and Daniel N. Robinson (London: Bloomsbury Academic, 2015), 10.

13. Preston, xviii.

14. Paul Lauritzen, "Stem Cells, Biotechnology, and Human Rights: Implications for a Posthuman Future," *Hastings Center Report* 35, no. 2 (2005), 25–26.

15. Table 2, Column 1.

16. Table 4, Column 1. Forty-three percent of those who read that a dozen NCs had been created somewhat or strongly agreed they support the creation of NCs, while 36.9 percent who read that scientists were creating a species were so agreeing. This is a small effect.

17. For analysis, I collapsed the "Strongly disagree" and "Somewhat disagree" categories.

18. See Table 6, Column 3. The effect is binary. Using the same method of examples used in the previous chapter, 75 percent of those who most agreed that species have their own purposes somewhat or strongly agreed that HBO research should continue, and 70 percent of those who least agree that species have their own purposes either somewhat or strongly agreed.

19. See Table 3, Column 3.

20. Michelle Wolkomir et al., "Substantive Religious Belief and Environmentalism," *Social Science Quarterly* 78, no. 1 (1997), 100; Wolkomir et al., "Denominational Subcultures of Environmentalism," *Review of Religious Research* 38, no. 4 (1997), 329.

21. Cronbach's alpha = .759. Higher values indicate stronger dominion views.

22. Table 6, Column 3.

23. See Table 3, Column 3.

24. Tables 3 and 6.

25. Peter Harrison, *The Territories of Science and Religion* (Chicago: University of Chicago Press, 2015).

26. Harrison, 154.

27. John H. Evans, *Morals Not Knowledge: Recasting the Contemporary U.S. Conflict between Religion and Science* (Berkeley: University of California Press, 2010), 63.

28. Stephen Jay Gould, *Rocks of Ages: Science and Religion in the Fullness of Life* (New York: Ballantine, 1999), 4.

29. Ian G. Barbour, *When Science Meets Religion: Enemies, Strangers, or Partners?* (San Francisco: Harper and Row, 2000), 2.

30. Elaine Howard Ecklund et al., *Secularity and Science: What Scientists around the World Really Think about Religion* (New York: Oxford University Press, 2019), 34.

31. Richard Dawkins, *The God Delusion* (New York: Bantam, 2006; reprint, Boston: Mariner Books, 2008).

32. Evans, *Morals Not Knowledge*, 69–76.

33. Robert G. Edwards, *Life before Birth: Reflections on the Embryo Debate* (New York: Basic Books, 1989), 165.

34. Quoted in *Man and His Future,* ed. G. E. W. Wolstenholme (New York: John Wiley, 1963), 372.

35. Tom Sorell, *Scientism: Philosophy and the Infatuation with Science* (London: Routledge, 2017).

36. Gregory R. Peterson, "Demarcation and the Scientistic Fallacy," *Zygon* 38, no. 4 (2003), 753.

37. Peterson, 752.

38. Stenmark, 29.

39. Robinson, 10.

40. Susan Haack, "Six Signs of Scientism," *Logos & Episteme* 3, no. 1 (February 1, 2012).

41. Robinson, 6–7.

42. Robinson, 3, 5.

43. Stenmark, 15.

44. Jan Slaby, "The New Science of Morality: A Bibliographic Review," *Hedgehog Review* 15, no. 1 (2013), 46.

45. Robinson, 14–15.

46. John H. Evans, *Contested Reproduction: Genetic Technologies, Religion, and Public Debate* (Chicago: University of Chicago Press, 2010), 54.

47. Robert Wuthnow and John H. Evans, *The Quiet Hand of God: Faith-Based Activism and the Public Role of Mainline Protestantism* (Berkeley: University of California Press, 2002).

48. John H. Evans, *Morals Not Knowledge*, Chap. 5.

49. Low-commitment Christians are Catholics or Protestants who do not fit into any of the other committed categories, primarily because they claim to be nonreligious or only slightly religious while still retaining their identity. I do not focus on this group because the impact of religion on them is presumably slight, and they only claim the identity because of long-past church participation.

50. These percentages for the Christian categories are lower than those cited in other sociological analyses because I am omitting those of minimal religious commitment (see footnote 50).

51. Ferenc Morton Szasz, *The Divided Mind of Protestant America, 1880-1930* (Tuscaloosa: University of Alabama Press, 1982), 34.

52. Christian Smith et al., *American Evangelicalism: Embattled and Thriving* (Chicago: University of Chicago Press, 2014), Chap. 1.

53. Robert Wuthnow, *The Restructuring of American Religion* (Princeton, NJ: Princeton University Press, 1988), Chap. 5.

54. Richard Owen, "Vatican Buries the Hatchet with Charles Darwin," *Times* (London), February 11, 2009.

55. Evans, *Morals Not Knowledge*, 124.

56. Evans, Chap. 7 of Morals not knowledge.

57. Evans and Hargittai, 6.

58. Phil Zuckerman, Luke W. Galen, and Frank L. Pasquale, *The Nonreligious: Understanding Secular People and Societies* (New York: Oxford University Press, 2016), Chap. 1.

59. Evans and Feng, 604–607.

60. See Table 6, Column 1. The percentage reported in the text is from a predicted value analysis.

61. Peter Wehner, "The Evangelical Church Is Breaking Apart," *Atlantic*, October 24, 2021.

62. For the full model, see Table 6, Column 3.

63. This analysis shows that Catholics are weakly more opposed.

64. See Table 3, Column 1.

65. I had a block of four questions that asked for agreement about four different conceptions of the human as described in Chapter 1 and created a measure of exclusively holding the Christian definition of a human. The four definitions of the human were: humans as defined by (1) their DNA, (2) their traits, (3) their ability to interact, or (4) being made in the image of God. I took the value on the image question (which was 1-5) and subtracted the average score on the other three questions. This resulted in a variable where the highest value is where the respondent strongly agreed with the image definition and strongly disagreed with all of the other three. The lowest value was where a respondent strongly agreed with all of the other three and strongly disagreed with the image one.

66. See Table 6, Column 3, and Table 3, Column 3.

67. Wolkomir et al., "Denominational Subcultures of Environmentalism," 326.

68. Robert Gifford and Andreas Nilsson, "Personal and Social Factors That Influence Pro-Environmental Concern and Behaviour: A Review," *International Journal of Psychology* 49, no. 3 (2014), 147.

69. Ronald Cole-Turner, *The New Genesis* (Louisville, KY: Westminster John Knox Press, 1993), 100.

70. This was reverse-coded so that the "oversee" end was given higher numbers. The case study I used to develop these survey questions was of evangelical scientists' views of human enhancement and transhumanism. Using this question, I found that those who those on the cocreator end of the spectrum were more likely to approve of both. See John H. Evans, "The Theological Debate over Human Enhancement: An Empirical Case Study of a Mediating Organization," *Zygon* 55, no. 3 (2020), 632.

71. See Table 6, Column 3.

72. Gary Dorrien, *Social Ethics in the Making: Interpreting an American Tradition* (Malden, MA: Wiley-Blackwell, 2010), 448–51.

73. This was reverse-coded so that those who see the Kingdom as independent of human activity received a "10." In the case study of evangelical scientists' views, I found that those who thought humans should create the Kingdom were more in favor of transhumanism but were not more or less in favor of human enhancement. See Evans, "Theological Debate," 632.

74. Robert Wuthnow, *Restructuring*, Chap. 5.

75. Gordon Gauchat, "Politicization of Science in the Public Sphere: A Study of Public Trust in the United States," 724; Gauchat, "The Political Context of Science in the United States, 1974–2010," *American Sociological Review* 77, no. 2 (April 2012), 167."

76. Marcus Mann and Cyrus Schleifer, "Love the Science, Hate the Scientists: Conservative Identity Protects Belief in Science and Undermines Trust in Scientists," *Social Forces* 99, no. 1 (August 5, 2020), 305.

77. Evans and Hargittai, "Who Doesn't Trust Fauci?" Therefore, I created indicators that the respondent identified with either the Democratic or Republican party or was independent. The views of Republicans and independents are compared to Democrats, 9.

78. See Table 3, Column 3, and Table 6, Column 3.

79. Anthony Dudo and John C. Besley, "Scientists' Prioritization of Communication Objectives for Public Engagement," *PLOS ONE* 11, no. 2 (February 25, 2016).

80. Molly J. Simis et al., "The Lure of Rationality: Why Does the Deficit Model Persist in Science Communication?" *Public Understanding of Science* 25, no. 4 (May 1, 2016), 402.

81. Matthew C. Nisbet and Dietram A. Scheufele, "What's Next for Science Communication? Promising Directions and Lingering Distractions," *American Journal of Botany* 96, no. 10 (Oct. 2009), 1767; Dietram A. Scheufele et al., "What We Know about Effective Public Engagement on CRISPR and Beyond," *Proceedings of the National Academy of Sciences* 118, no. 22 (June 1, 2021), 2.

82. See Table 6, Column 3.

83. See Table 3, Column 3.

84. The response categories for the trust question were (1) A great deal of confidence, (2) A fair amount of confidence, (3) Not too much confidence, and (4) No confidence at all. Since the first question had five

response categories and the second had four, I needed to make sure the two questions would have equal weight in the new index. I divided the response for each by five and four, respectively, to limit the range of each from zero to one. These were then added. The alpha for these two questions is .710.

85. Chronbach's Alpha = .835.

86. See the final column in Tables 3 and 6. These are fully standardized regression coefficients from the model in Column 3 of each table created using the listcoef function in Stata. I endorse all the usual cautions in interpreting fully standardized coefficients. See J. Scott Long and Jeremy Freese, *Regression Models for Categorical Dependent Variables Using Stata* (College Station, TX: Stata Press, 2014), 333.

Chapter 6

1. John H. Evans, "Can the Public Express Their Views or Say No through Public Engagement?," *Environmental Communication* 14, no. 7 (October 2, 2020), 882.

2. John H. Evans, *The History and Future of Bioethics: A Sociological View* (New York: Oxford University Press, 2012), Chap. 2.

3. Evans, *What Is a Human?*

4. Robert and Baylis, 10.

5. Stefan Timmermans and Rene Almeling, "Objectification, Standardization, and Commodification in Health Care: A Conceptual Readjustment," *Social Science & Medicine* 69, no. 1 (July 1, 2009), 22.

6. Joas, Chap. 2.

7. Gowan Dawson, *Darwin, Literature, and Victorian Respectability* (New York: Cambridge University Press, 2007).

8. Sara Shostak et al., "The Politics of the Gene: Social Status and Beliefs about Genetics for Individual Outcomes," *Social Psychology Quarterly* 72, no. 1 (March 1, 2009), 78.

9. Ronald L. Numbers, "Creationism in 20th-Century America," *Science* 218, no. 4572 (November 5, 1982), 538; Gould, 150–70.

10. Degeling, Irvine, and Kerridge, 268.

11. Keen, 25.

12. Monroe, 216.

13. Paul G. Bain, Jeroen Vaes, and Jacques Philippe Leyens, *Humanness and Dehumanization* (London: Psychology Press, 2013), 2.

14. Greely, "The Dilemma of Human Brain Surrogates: Scientific Opportunities, Ethical Concerns," in *Neuroscience and Law,* ed. Antonio D'Aloia and Maria Chiara Errigo (Cham, Switzerland: Springer, 2020), 389.

15. Greely, "Dilemma," 389.

16. Streiffer, 8.

17. Greely et al., 34.

18. This is the subtle-versus-explicit distinction in the psychology research. See Nick Haslam, "What Is Dehumanization?," in *Humanness and Dehumanization,* ed. Paul G. Bain, Jeroen Vaes, and Jacques-Philippe Leyens (New York: Psychology Press, 2014), 37.

19. Robert and Baylis, 6.

20. Leon R. Kass, "Organs for Sale? Propriety, Property, and the Price of Progress," *The Public Interest* 107 (Spring 1992), 82–83.

21. W. I. Thomas and D. S. Thomas, *The Child in America: Behavior Problems and Programs* (New York: Knopf, 1928), 572.

22. Wuthnow, *Producing the Sacred,* Chap. 1.

23. Calling the entities that violate categories "monsters," Smits writes that "Monster theory shows that monsters will come and go and may never be eradicated, since the dynamics of symbolic order dictate that a final exorcism of waste, danger, violence and monsters is impossible as a matter of principle. However, this does not force us into fatalism; motivated by an inherent need to order the symbolic universe, people untiringly attempt to remove smaller and larger anomalies. In that process, several ways or styles are open for domesticating monsters." See Smits, 500.

24. Leon R. Kass, "The Wisdom of Repugnance: Why We Should Ban the Cloning of Humans," *Valparaiso University Law Review* 32, no. 2 (Spring 1998), 687.

25. Morriss, 288.

26. Morrison and de Saille, 6.

27. Smits, 500–502. There are actually four, but only three are useful for analyzing an NC. The one that is not particularly useful for us is "embracing," in which we turn the entity into a miracle or saint.

28. Streiffer, 14.

29. Nik Brown, 327.

30. "The designation of a thing, event, deed, or person as 'dirty' and 'out of place' is rarely an unalterable verdict as 'most pollutions have a very simple remedy for undoing their effects. There are rites of reversing, untying, burying, washing, erasing, fumigating, and so on'" Morrison and de Saille, 6.

31. Douglas, *Purity and Danger,* 40.

32. Maartje Schermer, 223.

33. Megan Molteni, "How 'Self-Limiting' Mosquitos Can Help Eradicate Malaria," *Wired,* June 21, 2018, www.wired.com/story/oxitec-gates-self-limiting-mosquitos.

34. Evans, *What Is a Human?*

35. Wibren van der Burg, "The Slippery Slope Argument," *Ethics* 102, no. 1 (1991); David Albert Jones, "Is There a Logical Slippery Slope from Voluntary to Nonvoluntary Euthanasia?," *Kennedy Institute of Ethics Journal* 21, no. 4 (2011), 384; Douglas Walton, *The Slippery Slope Argument* (New York: Oxford University Press, 1992).

36. Evans, *Human Gene Editing Debate,* Chap. 1.

37. Evans, 13–15.

38. Mario J. Rizzo and Douglas Glen Whitman, "Little Brother Is Watching You: New Paternalism on the Slippery Slopes," *Arizona Law Review* 51 (2009), 691.

39. Rizzo and Whitman, 691.

40. While a working compromise on locations on the slope subject to continuity vagueness do occur (as on the subject of driving age), this requires consensus that both extremes are unacceptable to all.

41. Rizzo and Whitman, 691.

42. Bok, 9.

43. Jones, 385; Rizzo and Whitman, 738.

44. Evans, *The Human Gene Editing Debate,* Chaps. 2 and 3.

45. Insoo Hyun, Amy Wilkerson, and Josephine Johnston, "Embryology Policy: Revisit the 14-Day Rule," *Nature* 533, no. 7602 (May 2016).

46. Rebecca M. Marton and Sergiu P. Paşca, "Organoid and Assembloid Technologies for Investigating Cellular Crosstalk in Human Brain Development and Disease," *Trends in Cell Biology* 30, no. 2 (February 1, 2020), 133.

47. Cyriel M. A. Pennartz, Michele Farisco, and Kathinka Evers, "Indicators and Criteria of Consciousness in Animals and Intelligent Machines: An Inside-Out Approach," *Frontiers in Systems Neuroscience* 13 (July 16, 2019); Anil K. Seth, "Consciousness: The Last 50 Years (and the Next)," *Brain and Neuroscience Advances* 2 (January 2018).

48. Jack-Yves Deschamps et al., "History of Xenotransplantation," *Xenotransplantation* 12, no. 2 (2005), 92.

49. Deschamps et al., 92, 95.

50. De Los Angeles, see n.26 in Chapter 1, 333. Kok Hooi Yap et al., "Aortic Valve Replacement: Is Porcine or Bovine Valve Better?" *Interactive CardioVascular and Thoracic Surgery* 16, no. 3 (March 1, 2013), 361.

51. Leonard D. Shultz, Fumihiko Ishikawa, and Dale L. Greiner, "Humanized Mice in Translational Biomedical Research," *Nature Reviews Immunology* 7, no. 2 (February 2007), 119.

52. Markus Grompe and Stephen Strom, "Mice with Human Livers," *Gastroenterology* 145, no. 6 (December 1, 2013).

53. Stephen R. Latham, "U.S. Law and Animal Experimentation: A Critical Primer," *Hastings Center Report* 35, no. 2 (2005), S35.

54. Quoted in Greely, "Human Brain Surrogates Research: The Onrushing Ethical Dilemma, *American Journal of Bioethics* 21, no. 1 (2021), 40.

Methodological Appendix

1. Alexander Coppock and Oliver A. McClellan, "Validating the Demographic, Political, Psychological, and Experimental Results Obtained from a New Source of Online Survey Respondents," *Research & Politics* 6, no. 1 (January 1, 2019); Seth J. Hill and Gregory A. Huber, "On the Meaning of Survey Reports of Roll-Call 'Votes,'" *American Journal of Political Science* 63, no. 3 (July 2019), 615.

2. Kyle Peyton, Gregory A. Huber, and Alexander Coppock, "The Generalizability of Online Experiments Conducted during the COVID-19 Pandemic," *Journal of Experimental Political Science* 9, no. 3 (Winter 2022), 1.

3. Ansolabehere and Schaffner, 89.

4. Adam J. Berinsky et al., "Using Screeners to Measure Respondent Attention on Self-Administered Surveys: Which Items and How Many?" *Political Science Research and Methods* 9, no. 2 (April 2021); Adam J. Berinsky, Michele F. Margolis, and Michael W. Sances, "Separating the Shirkers from the Workers? Making Sure Respondents Pay Attention on Self-Administered Surveys," *American Journal of Political Science* 58, no. 3 (2014).

5. Adam J. Berinsky, Michele F. Margolis, and Michael W. Sances, "Can We Turn Shirkers into Workers?," *Journal of Experimental Social Psychology* 66 (September 1, 2016), 22.

6. Peyton, Huber, and Coppock, 11.

7. Berinsky, Margolis, and Sances, "Can We Turn Shirkers into Workers?," 20.

8. Diana C. Mutz, *Population-Based Survey Experiments* (Princeton, NJ: Princeton University Press, 2011), 87. For the HBO vignette I stated, "The description said that it is the electrical activity of an organoid that could lead to cures for Alzheimer's disease. a) True; b) False. For the NC

vignette I stated: "The description said that the modification in the animal will exactly match a human disease: a) True; b) False."

9. Berinsky, Margolis, and Sances, "Separating the Shirkers from the Workers?"

10. Peyton, Huber, and Coppock, 9.

11. Mutz, 88–89.

12. Peter M. Aronow, Jonathon Baron, and Lauren Pinson, "A Note on Dropping Experimental Subjects Who Fail a Manipulation Check," *Political Analysis* 27 (2019), 576.

13. For gender: https://www.census.gov/quickfacts/fact/table/US/LFE046 219; for age: https://data.census.gov/cedsci/table?q=age&tid=ACSST1Y2 019.S0101; for education: https://www.census.gov/data/tables/2020/ demo/educational-attainment/cps-detailed-tables.html; for race/ethnicity: https://www.census.gov/quickfacts/fact/table/US/LFE046219. Race and ethnicity are measured in many different ways, and the census data is based on sole-identities. I weighted on "White," "Black," "Hispanic," and "Other."

14. Andrew Mercer, Arnold Lau, and Courtney Kennedy, "For Weighting Online Opt-In Samples, What Matters Most?" January 26, 2018, 4, https:// policycommons.net/artifacts/617484/for-weighting-online-opt-in-samp les-what-matters-most/1598296/ (accessed July 20, 2023).

15. Hill and Huber, Appendix D.

16. In my data, Catholics are those who selected "Catholic" on the basic religion question, and Protestants are those who selected either "Protestant" or "Just Christian." In the GSS, Protestants are those who selected "Protestant" or "Christian." The GSS Bible question, replicated in my survey, asks "Which of these statements comes closest to describing your feelings about the Bible? The Bible is the actual word of God and is to be taken literally, word for word; The Bible is the inspired word of God but not everything in it should be taken literally, word for word; The Bible is an ancient book of fables, legends, history, and moral precepts recorded by man."

17. Richard Valliant and Jill A. Dever, *Survey Weights: A Step-by-Step Guide to Calculation* (College Station, TX: Stata Press, 2018), 59.

18. Valliant and Dever, 59.

19. Elaine Howard Ecklund and Christopher P. Scheitle, *Religion vs. Science: What Religious People Really Think* (New York: Oxford University Press, 2018), 158.

20. Kevin D. Dougherty, Byron R. Johnson, and Edward C. Polson, "Recovering the Lost: Remeasuring U.S. Religious Affiliation," *Journal for the Scientific Study of Religion* 46, no. 4 (December 7, 2007), 485.

21. Dougherty, Johnson, and Polson, 483.

22. Derek Lehman and Darren E. Sherkat, "Measuring Religious Identification in the United States," *Journal for the Scientific Study of Religion* 57, no. 4 (December 2018), 786.

23. David Johnson, Christopher Scheitle, and Elaine Ecklund, "Individual Religiosity and Orientation towards Science: Reformulating Relationships," *Sociological Science* 2 (2015), 114.

24. R. D. Woodberry et al., "The Measure of American Religious Traditions: Theoretical and Measurement Considerations," *Social Forces* 91, no. 1 (September 1, 2012), 69.

25. Lehman and Sherkat, 783.

26. Evans and Hargittai, "Who Doesn't Trust Fauci?" 11.

27. Donald E. Miller, *Reinventing American Protestantism: Christianity in the New Millennium* (Berkeley: University of California Press, 1997), 129.

28. The differences I mention are statistically significantly different at the p<.05 level or less.

29. On a five-point educational achievement scale, identity-rejecting Protestants score 2.7, conservative Protestants 3.1, and liberal Protestants 3.6. The percent women in the three groups above, respectively, 67, 54, and 51. The average ages are, respectively, 50.5, 57.1, and 62.7. On a seven-point income scale, the respective averages for the three groups are 2.2, 2.7, and 2.9.

30. The metric for this scale is not intuitive, but it ranges from -4 to 4. Identity-rejecting Protestants score .33 and conservative Protestants score .55.

31. Smith, Christian and Melinda Lundquist Denton, *Soul Searching: The religious and spiritual lives of American teenagers,* (New York: Oxford University Press, 2005), 162–70.

32. J. Scott Long and Jeremy Freese, *Regression Models for Categorical Dependent Variables Using Stata,* 3rd ed. (College Station, TX: Stata Press, 2014), 326.

33. Long and Freese, 331.

34. Richard Williams, "Generalized Ordered Logit/Partial Proportional Odds Models for Ordinal Dependent Variables," *Stata Journal* 6, no. 1 (February 1, 2006), 58.

35. Richard Williams, "Understanding and Interpreting Generalized Ordered Logit Models," *Journal of Mathematical Sociology* 40, no. 1 (January 2, 2016), 15.

36. John V. Kane and Jason Barabas, "No Harm in Checking: Using Factual Manipulation Checks to Assess Attentiveness in Experiments," *American Journal of Political Science* 63, no. 1 (2019), 238.

37. Kane and Barabas, 248.

38. Mutz, 85.

39. An ordered logistic model predicting the six-point response by the dichotomous measure of the new species vs. dozen factor showed that the new species level predicted more likelihood of seeing the entities (p=.044).

40. Peter M. Aronow, Jonathon Baron, and Lauren Pinson, "A Note on Dropping Experimental Subjects Who Fail a Manipulation Check," *Political Analysis* 27 (2019), 573.

Works Cited

Academy of Medical Sciences. *Animals Containing Human Material*. London: Academy of Medical Sciences, 2011.

Altman, Claire E. "Age, Period, and Cohort Effects." In *Encyclopedia of Migration*, edited by Frank D. Bean and Susan K. Brown, 1–4. Dordrecht: Springer Netherlands, 2015.

Ansolabehere, Stephen, and Brian F. Schaffner. "Taking the Study of Political Behavior Online." In *The Oxford Handbook of Polling and Survey Methods*, edited by Lonna Rae Atkeson and R. Michael Alvarez. New York: Oxford University Press, 2018.

Aronow, Peter M, Jonathon Baron, and Lauren Pinson. "A Note on Dropping Experimental Subjects Who Fail a Manipulation Check." *Political Analysis* 27 (2019): 572–89.

Atkinson, Stuart. "Neural Blastocyst Complementation: A New Means to Study Brain Development in Mouse." *Stem Cells Portal*, January 21, 2019, https://stemcellsportal.com/article-scans/neural-blastocyst-complementat ion-new-means-study-brain-development-mouse.

Bain, Paul G., Jeroen Vaes, and Jacques Philippe Leyens. *Humanness and Dehumanization*. London: Psychology Press, 2013.

Barbour, Ian G. *When Science Meets Religion: Enemies, Strangers, or Partners?* San Francisco: Harper and Row, 2000.

Begley, Sharon. "Tiny Human Brain Organoids Implanted into Rodents, Triggering Ethical Concerns." *STAT*, November 6, 2017, https://www.statn ews.com/2017/11/06/human-brain-organoids-ethics/.

Berger, Peter L., and Thomas Luckmann. *The Social Construction of Reality: A Treatise in the Sociology of Knowledge*. New York: Anchor, 1967.

Berinsky, Adam J., Michele F. Margolis, and Michael W. Sances. "Separating the Shirkers from the Workers? Making Sure Respondents Pay Attention on Self-Administered Surveys." *American Journal of Political Science* 58, no. 3 (2014): 739–53.

Berinsky, Adam J., Michele F. Margolis, and Michael W. Sances. "Can We Turn Shirkers into Workers?" *Journal of Experimental Social Psychology* 66 (September 1, 2016): 20–28.

Berinsky, Adam J., Michele F. Margolis, Michael W. Sances, and Christopher Warshaw. "Using Screeners to Measure Respondent Attention on Self-Administered Surveys: Which Items and How Many?" *Political Science Research and Methods* 9, no. 2 (April 2021): 430–37.

Bok, Sissela. "The Leading Edge of the Wedge." *The Hastings Center Report* 1, no. 3 (December 1971): 9–11.

Bolo, Isabel, Ben Curran Wills, and Karen J. Maschke. "Public Attitudes toward Human-Animal Chimera Research May Be More Complicated Than They Appear." *Stem Cell Reports* 16, no. 2 (February 2021): 225–26.

Brown, Nik. "Xenotransplantation: Normalizing Disgust." *Science as Culture* 8, no. 3 (September 1999): 327–55.

Brown, Warren S. "Cognitive Contributions to Soul." In *Whatever Happened to the Soul? Scientific and Theological Portraits of Human Nature*, edited by Warren S. Brown, Nancey Murphy, and H. Newton Malony, 99–125. Minneapolis, MN: Augsburg Press, 1998.

Burg, Wibren van der. "The Slippery Slope Argument." *Ethics* 102, no. 1 (1991): 42–65.

Chalmers, David J. "The Matrix as Metaphysics." In *Philosophers Explore the Matrix*, edited by Christopher Grau, 132–76. New York: Oxford University Press, 2005.

Chang, Amelia N., Zhuoyi Liang, Hai-Qiang Dai, Aimee M. Chapdelaine-Williams, Nick Andrews, Roderick T. Bronson, Bjoern Schwer, and Frederick W. Alt. "Neural Blastocyst Complementation Enables Mouse Forebrain Organogenesis." *Nature* 563, no. 7729 (November 2018): 126–30.

Chen, H. Isaac, John A. Wolf, Rachel Blue, Mingyan Maggie Song, Jonathan D. Moreno, Guo-li Ming, and Hongjun Song. "Transplantation of Human Brain Organoids: Revisiting the Science and Ethics of Brain Chimeras." *Cell Stem Cell* 25, no. 4 (October 2019): 462–72.

Coghlan, Andy. "The Smart Mouse with the Half-Human Brain." *New Scientist*, December 1, 2014, https://www.newscientist.com/article/dn26639-the-smart-mouse-with-the-half-human-brain/.

Cole-Turner, Ronald. *The New Genesis: Theology and the Genetic Revolution*. Louisville, KY: Westminster John Knox Press, 1993.

Collins, Debbie. *Cognitive Interviewing Practice*. Los Angeles: SAGE Publications, 2014.

Committee on Ethical, Legal, and Regulatory Issues Associated with Neural Chimeras and Organoids, Committee on Science, Technology, and Law, Policy and Global Affairs, and National Academies of Sciences, Engineering, and Medicine. *The Emerging Field of Human Neural Organoids, Transplants, and Chimeras: Science, Ethics, and Governance*. Washington, DC: National Academies Press, 2021.

Committee on Gene Drive Research in Non-Human Organisms: Recommendations for Responsible Conduct, Board on Life Sciences, Division on Earth and Life Studies, and National Academies of Sciences, Engineering, and Medicine. *Gene Drives on the Horizon: Advancing Science, Navigating Uncertainty, and Aligning Research with Public Values*. Washington, DC: National Academies Press, 2016.

Comosy, Charles, and Susan Kopp. "The Use of Non-Human Animals in Biomedical Research: Can Moral Theology Fill the Gap?" *Journal of Moral Theology* 3, no. 2 (2014): 54–71.

Coppock, Alexander, and Oliver A. McClellan. "Validating the Demographic, Political, Psychological, and Experimental Results Obtained from a New Source of Online Survey Respondents." *Research & Politics* 6, no. 1 (January 1, 2019): 1–14.

Crane, Andrew T., Francis X. Shen, Jennifer L. Brown, Warren Cormack, Mercedes Ruiz-Estevez, Joseph P. Voth, Tsutomu Sawai, Taichi Hatta, Misao Fujita, and Walter C. Low. "The American Public Is Ready to Accept Human-Animal Chimera Research." *Stem Cell Reports* 15, no. 4 (October 13, 2020): 804–10.

Dawkins, Richard. *The God Delusion*. New York: Bantam, 2006. Reprint, Boston: Mariner Books, 2008.

Dawson, Gowan. *Darwin, Literature, and Victorian Respectability*. New York: Cambridge University Press, 2007.

De Cruz, Helen, and Johan De Smedt. "The Role of Intuitive Ontologies in Scientific Understanding—the Case of Human Evolution." *Biology & Philosophy* 22, no. 3 (June 2007): 351–68.

Degeling, Chris, Rob Irvine, and Ian Kerridge. "Faith-Based Perspectives on the Use of Chimeric Organisms for Medical Research." *Transgenic Research* 23, no. 2 (April 2014): 265–79.

De Los Angeles, Alejandro, Nam Pho, and D. Eugene Redmond, Jr. "Generating Human Organs via Interspecies Chimera Formation: Advances and Barriers." *Yale Journal of Biology and Medicine* 91, no. 3 (September 21, 2018): 333–42.

Deschamps, Jack-Yves, Françoise A. Roux, Pierre Saï, and Edouard Gouin. "History of Xenotransplantation." *Xenotransplantation* 12, no. 2 (2005): 91–109.

Dolan, Brian. "Soul Searching: A Brief History of the Mind/Body Debate in the Neurosciences." *Neurosurgical Focus* 23, no. 1 (July 2007): 1–7.

Dorrien, Gary. *Social Ethics in the Making: Interpreting an American Tradition*. Malden, MA: Wiley-Blackwell, 2011.

Dougherty, Kevin D., Byron R. Johnson, and Edward C. Polson. "Recovering the Lost: Remeasuring U.S. Religious Affiliation." *Journal for the Scientific Study of Religion* 46, no. 4 (December 7, 2007): 483–99.

Douglas, Mary. *Natural Symbols: Explorations in Cosmology*. Routledge, 2002.

Douglas, Mary. *Purity and Danger: An Analysis of the Concepts of Pollution and Taboo*. London: Routledge and Kegan Paul, 1966. Reprint, London: Routledge, 1984.

Dudo, Anthony, and John C. Besley. "Scientists' Prioritization of Communication Objectives for Public Engagement." *PLOS ONE* 11, no. 2 (February 25, 2016), https://journals.plos.org/plosone/article?id=10.1371/journal.pone.0148867.

Duschinsky, Robbie. "Introduction." In *Purity and Danger Now: New Perspectives*, edited by Robbie Duschinsky, Simone Schnall, and Daniel H. Weiss, 1–19. New York: Routledge, 2016.

Ecklund, Elaine Howard, and Christopher P. Scheitle. *Religion vs. Science: What Religious People Really Think*. New York: Oxford University Press, 2018.

Ecklund, Elaine Howard, David R. Johnson, Brandon Vaidyanathan, Kirstin R. W. Matthews, Steven W. Lewis, Robert A. Thomson, Jr, and Di Di. *Secularity and Science: What Scientists around the World Really Think about Religion*. New York: Oxford University Press, 2019.

Edwards, Robert G. *Life before Birth: Reflections on the Embryo Debate*. New York: Basic Books, 1989.

Engelke, Matthew. *How to Think Like an Anthropologist*. Princeton, NJ: Princeton University Press, 2019.

Evans, John H. *Contested Reproduction: Genetic Technologies, Religion, and Public Debate*. Chicago: University of Chicago Press, 2010.

Evans, John H. *The History and Future of Bioethics: A Sociological View*. New York: Oxford University Press, 2012.

Evans, John H. *What Is a Human?: What the Answers Mean for Human Rights*. New York: Oxford University Press, 2016.

Evans, John H. *Morals Not Knowledge: Recasting the Contemporary U.S. Conflict between Religion and Science*. Berkeley: University of California Press, 2018.

Evans, John H. *The Human Gene Editing Debate*. New York: Oxford University Press, 2020.

Evans, John H. "Can the Public Express Their Views or Say No through Public Engagement?" *Environmental Communication* 14, no. 7 (October 2, 2020): 881–85.

Evans, John H. "The Theological Debate over Human Enhancement: An Empirical Case Study of a Mediating Organization." *Zygon* 55, no. 3 (2020): 615–37.

Evans, John H., and Justin Feng. "Conservative Protestantism and Skepticism of Scientists Studying Climate Change." *Climatic Change* 121, no. 4 (December 1, 2013): 595–608.

Evans, John H., and Eszter Hargittai. "Who Doesn't Trust Fauci? The Public's Belief in the Expertise and Shared Values of Scientists in the COVID-19 Pandemic." *Socius* 6 (January 1, 2020): 2378023120947337, https://journals.sagepub.com/doi/full/10.1177/2378023120947337.

Farahany, Nita A., Henry T. Greely, Steven Hyman, Christof Koch, Christine Grady, Sergiu P. Paşca, Nenad Sestan et al. "The Ethics of Experimenting with Human Brain Tissue." *Nature* 556, no. 7702 (April 2018): 429–32.

Feng, Guoping, Frances E. Jensen, Henry T. Greely, Hideyuki Okano, Stefan Treue, Angela C. Roberts, James G. Fox et al. "Opportunities and Limitations of Genetically Modified Nonhuman Primate Models for

Neuroscience Research." *Proceedings of the National Academy of Sciences* 117, no. 39 (August 19, 2020), 24022–24031.

Fox, Renee C., and Judith P. Swazey. *Spare Parts: Organ Replacement in American Society*. New York: Oxford University Press, 1992.

Funk, Cary, and Meg Hefferon. "Most Americans Accept Genetic Engineering of Animals That Benefits Human Health, but Many Oppose Other Uses." Washington, DC: Pew Research Center, 2018, https://www.pewresearch.org/science/2018/08/16/most-americans-accept-genetic-engineering-of-animals-that-benefits-human-health-but-many-oppose-other-uses/.

Gabriel, Elke, Walid Albanna, Giovanni Pasquini, Anand Ramani, Natasia Josipovic, Aruljothi Mariappan, Friedrich Schinzel et al. "Human Brain Organoids Assemble Functionally Integrated Bilateral Optic Vesicles," *Cell Stem Cell* 28 (October 7, 2021): 1740–57.

Gauchat, Gordon. "Politicization of Science in the Public Sphere: A Study of Public Trust in the United States, 1974 to 2010." *American Sociological Review* 77, no. 2 (April 2012): 167–87.

Gauchat, Gordon. "The Political Context of Science in the United States: Public Acceptance of Evidence-Based Policy and Science Funding." *Social Forces* 94, no. 2 (December 2015): 723–46.

Gauchat, Gordon, Timothy O'Brien, and Oriol Mirosa. "The Legitimacy of Environmental Scientists in the Public Sphere." *Climatic Change* 143, nos. 3–4 (August 2017): 297–306.

Gifford, Robert, and Andreas Nilsson. "Personal and Social Factors That Influence Pro-Environmental Concern and Behaviour: A Review." *International Journal of Psychology* 49, no. 3 (2014): 141–57.

Gould, Stephen Jay. *Rocks of Ages: Science and Religion in the Fullness of Life*. New York: Ballantine, 1999.

Greely, Henry T. "Human Brain Surrogates Research: The Onrushing Ethical Dilemma." *American Journal of Bioethics* 21, no. 1 (2021): 34–45.

Greely, Henry T. "Human/Nonhuman Chimeras: Assessing the Issues." In *The Oxford Handbook of Animal Ethics*, edited by Tom R. Beauchamp and R. G. Frey, 671–700. New York: Oxford University Press, 2011.

Greely, Henry T. "The Dilemma of Human Brain Surrogates: Scientific Opportunities, Ethical Concerns." In *Neuroscience and Law*, edited by Antonio D'Aloia and Maria Chiara Errigo, 371–99. Cham, Switzerland: Springer International Publishing, 2020.

Greely, Henry T., Mildred K. Cho, Linda F. Hogle, and Debra M. Satz. "Thinking about the Human Neuron Mouse." *American Journal of Bioethics* 7, no. 5 (May 10, 2007): 27–40.

Greely, Henry T., and Nita A. Farahany. "Advancing the Ethical Dialogue about Monkey/Human Chimeric Embryos." *Cell* 184, no. 8 (April 15, 2021): 1962–63.

Grompe, Markus, and Stephen Strom. "Mice With Human Livers." *Gastroenterology* 145, no. 6 (December 1, 2013): 1209–14.

Guo, Kun, David Tunnicliffe, and Hettie Roebuck. "Human Spontaneous Gaze Patterns in Viewing of Faces of Different Species." *Perception* 39, no. 4 (April 1, 2010): 533–42.

Haack, Susan. "Six Signs of Scientism." *Logos & Episteme* 3, no. 1 (February 1, 2012): 75–95.

Haddow, Gill. "Animal, Mechanical, and Me: Organ Transplantation and the Ambiguity of Embodiment." In *The Oxford Handbook of the Sociology of Body and Embodiment*, edited by Natalie Boero and Katherine Mason, Chapter 10. New York: Oxford University Press, 2020.

Haddow, Gill. *Embodiment and Everyday Cyborgs: Technologies That Alter Subjectivity*. Manchester: Manchester University Press, 2021.

Hagelin, J. "Public Opinion Surveys about Xenotransplantation." *Xenotransplantation* 11, no. 6 (2004): 551–58.

Han, Xiaoning, Michael Chen, Fushun Wang, Martha Windrem, Su Wang, Steven Shanz, Qiwu Xu et al. "Forebrain Engraftment by Human Glial Progenitor Cells Enhances Synaptic Plasticity and Learning in Adult Mice." *Cell Stem Cell* 12, no. 3 (March 7, 2013): 342–53.

Harrison, Peter. *The Territories of Science and Religion*. Chicago: University of Chicago Press, 2015.

Haslam, Nick. "What Is Dehumanization?" In *Humanness and Dehumanization*, edited by Paul G. Bain, Jeroen Vaes, and Jacques-Philippe Leyens, 34–48. New York: Psychology Press, 2014.

Heinlein, Robert A. "Jerry Was a Man." In *Assignment in Eternity*, 215–42. Reading, PA: Fantasy Press, 1953. Reprint, Riverdale, Canada: Baen Books, 2012.

Hill, Seth J., and Gregory A. Huber. "On the Meaning of Survey Reports of Roll-Call 'Votes.'" *American Journal of Political Science* 63, no. 3 (July 2019): 611–25.

Hinterberger, Amy. "Regulating Estrangement: Human–Animal Chimeras in Postgenomic Biology." *Science, Technology, & Human Values* 45, no. 6 (November 2020): 1065–86.

Hübner, Dietmar. "Human-Animal Chimeras and Hybrids: An Ethical Paradox behind Moral Confusion?" *Journal of Medicine and Philosophy* 43, no. 2 (March 13, 2018): 187–210.

Hyun, Insoo, J. C. Scharf-Deering, and Jeantine E. Lunshof. "Ethical Issues Related to Brain Organoid Research." *Brain Research* 1732 (April 1, 2020): 146653.

Hyun, Insoo, Amy Wilkerson, and Josephine Johnston. "Embryology Policy: Revisit the 14-Day Rule." *Nature* 533, no. 7602 (May 2016): 169–71.

Ikels, Charlotte. "The Anthropology of Organ Transplantation." *Annual Review of Anthropology* 42, no. 1 (2013): 89–102.

Joas, Hans. *The Sacredness of the Person: A New Genealogy of Human Rights.* Washington, DC: Georgetown University Press, 2013.

Johnson, David, Christopher Scheitle, and Elaine Ecklund. "Individual Religiosity and Orientation towards Science: Reformulating Relationships." *Sociological Science* 2 (2015): 106–24.

Jones, David Albert. "Is There a Logical Slippery Slope from Voluntary to Nonvoluntary Euthanasia?" *Kennedy Institute of Ethics Journal* 21, no. 4 (2011): 379–404.

Kaebnick, Gregory E. *Humans in Nature: The World As We Find It and the World As We Create It.* New York: Oxford University Press, 2013.

Kane, John V., and Jason Barabas. "No Harm in Checking: Using Factual Manipulation Checks to Assess Attentiveness in Experiments." *American Journal of Political Science* 63, no. 1 (2019): 234–49.

Karpowicz, Phillip, Cynthia B Cohen, and Derek van der Kooy. "It Is Ethical to Transplant Human Stem Cells into Nonhuman Embryos." *Nature Medicine* 10, no. 4 (April 2004): 331–35.

Kass, Leon R. "Organs for Sale? Propriety, Property, and the Price of Progress." *Public Interest* 107 (Spring 1992): 65–86.

Kass, Leon R. "The Wisdom of Repugnance: Why We Should Ban the Cloning of Humans." *Valparaiso University Law Review* 32, no. 2 (Spring 1998): 679–705.

Katsyri, Jari, Klaus Forger, Meeri Makarainen, and Tapio Takala. "A Review of Empirical Evidence on Different Uncanny Valley Hypotheses: Support for Perceptual Mismatch as One Road to the Valley of Eeriness." *Frontiers in Psychology* 6 (April 10, 2015): 1–16.

Keen, Sam. *Faces of the Enemy: Reflections of the Hostile Imagination.* San Francisco: Harper and Row, 1986.

King, Anthony. "The Search for Better Animal Models of Alzheimer's Disease." *Nature* 559, no. 7715 (July 25, 2018): S13–15.

King, Mike R., Maja I. Whitaker, and D. Gareth Jones. "I See Dead People: Insights from the Humanities into the Nature of Plastinated Cadavers." *Journal of Medical Humanities* 35, no. 4 (December 2014): 361–76.

Klitzman, Robert. "How Artistic Representation Can Inform Current Debates about Chimeras." *Journal of Medical Humanities* 42, no. 3 (September 1, 2021): 337–43.

Knoppers, Bartha Maria, and Henry T. Greely. "Biotechnologies Nibbling at the Legal 'Human.'" *Science* 366, no. 6472 (December 20, 2019): 1455–57.

Latham, Stephen R. "U.S. Law and Animal Experimentation: A Critical Primer." *Hastings Center Report* 42, no. s1 (2012): S35–39.

Lauritzen, Paul. "Stem Cells, Biotechnology, and Human Rights: Implications for a Posthuman Future." *Hastings Center Report* 35, no. 2 (2005): 25–33.

Lavazza, Andrea, and Marcello Massimini. "Cerebral Organoids: Ethical Issues and Consciousness Assessment." *Journal of Medical Ethics* 44, no. 9 (September 2018): 606–10.

Lebacqz, Karen. "Alien Dignity: The Legacy of Helmut Thielicke for Bioethics." In *Religion and Medical Ethics: Looking Back, Looking Forward*, edited by Allen Verhey, 44–60. Grand Rapids, MI: Eerdmans, 1996.

Lee, Sandra S.-J., Mildred K. Cho, Stephanie A. Kraft, Nina Varsava, Katie Gillespie, Kelly E. Ormond, Benjamin S. Wilfond, and David Magnus. "'I Don't Want to Be Henrietta Lacks': Diverse Patient Perspectives on Donating Biospecimens for Precision Medicine Research." *Genetics in Medicine* 21, no. 1 (January 2019): 107–13.

Lehman, Derek, and Darren E. Sherkat. "Measuring Religious Identification in the United States." *Journal for the Scientific Study of Religion* 57, no. 4 (December 2018): 779–94.

Lindee, M. Susan, and Dorothy Nelkin. *The DNA Mystique: The Gene as a Cultural Icon*. New York: W. H. Freeman, 1995.

Liu, Zhen, Xiao Li, Jun-Tao Zhang, Yi-Jun Cai, Tian-Lin Cheng, Cheng Cheng, Yan Wang et al. "Autism-like Behaviours and Germline Transmission in Transgenic Monkeys Overexpressing MeCP2." *Nature* 530, no. 7588 (February 2016): 98–102.

Loike, John D. "Opinion: Should Human-Animal Chimeras Be Granted 'Personhood'?" *Scientist*, May 23, 2018, https://www.the-scientist.com/news-opinion/opinion-should-human-animal-chimeras-be-granted-personhood-36664..

Long, J. Scott, and Jeremy Freese. *Regression Models for Categorical Dependent Variables Using Stata*. 3rd ed. College Station, TX: Stata Press, 2014.

Macnaghten, Phil. "Animals in Their Nature: A Case Study on Public Attitudes to Animals, Genetic Modification and 'Nature.'" *Sociology* 38, no. 3 (2004): 533–51.

Mann, Marcus, and Cyrus Schleifer. "Love the Science, Hate the Scientists: Conservative Identity Protects Belief in Science and Undermines Trust in Scientists." *Social Forces* 99, no. 1 (August 5, 2020): 305–32.

Marton, Rebecca M., and Sergiu P. Paşca. "Organoid and Assembloid Technologies for Investigating Cellular Crosstalk in Human Brain Development and Disease." *Trends in Cell Biology* 30, no. 2 (February 1, 2020): 133–43.

Mercer, Andrew, Arnold Lau, and Courtney Kennedy. "For Weighting Online Opt-In Samples, What Matters Most?," 2018, 4. https://policycommons.net/artifacts/617484/for-weighting-online-opt-in-samples-what-matters-most/1598296/ on 20 Jul 2023. CID: 20.500.12592/1zfbv0.

Miller, Donald E. *Reinventing American Protestantism: Christianity in the New Millennium*. Berkeley: University of California Press, 1997.

Molteni, Megan. "How 'Self-Limiting' Mosquitos Can Help Eradicate Malaria." *Wired*, June 21, 2018, https://www.wired.com/story/oxitec-gates-self-limiting-mosquitoes/.

Monroe, Kristen Renwick. "Review Essay: The Psychology of Genocide." *Ethics & International Affairs* 9 (March 1995): 215–39.

Morrison, Michael, and Stevienna de Saille. "CRISPR in Context: Towards a Socially Responsible Debate on Embryo Editing." *Palgrave Communications* 5, no. 1 (December 2019): 1–9.

Morriss, Peter. "Blurred Boundaries." *Inquiry* 40, no. 3 (September 1997): 259–89.

Mutz, Diana C. *Population-Based Survey Experiments*. Princeton, NJ: Princeton University Press, 2011.

Nisbet, Matthew C., and Dietram A. Scheufele. "What's Next for Science Communication? Promising Directions and Lingering Distractions." *American Journal of Botany* 96, no. 10 (October 2009): 1767–78.

Numbers, Ronald L. "Creationism in 20th-Century America." *Science* 218, no. 4572 (November 5, 1982): 538–44.

Ormandy, Elisabeth H., and Catherine A. Schuppli. "Public Attitudes toward Animal Research: A Review." *Animals* 4, no. 3 (September 2014): 391–408.

Owen, Richard. "Vatican Buries the Hatchet with Charles Darwin." *Times* (London). February 11, 2009.

Parry, Sarah. "Interspecies Entities and the Politics of Nature." *Sociological Review* 58, no. 1 suppl. (May 2010): 113–29.

Pennartz, Cyriel M. A., Michele Farisco, and Kathinka Evers. "Indicators and Criteria of Consciousness in Animals and Intelligent Machines: An Inside-Out Approach." *Frontiers in Systems Neuroscience* 13 (July 16, 2019): 1–23.

Peterson, Gregory R. "Demarcation and the Scientistic Fallacy." *Zygon* 38, no. 4 (2003): 751–61.

Peyton, Kyle, Gregory A. Huber, and Alexander Coppock. "The Generalizability of Online Experiments Conducted during the COVID-19 Pandemic." *Journal of Experimental Political Science* 9, no. 3 (Winter 2022): 379–94.

Plantinga, Alvin. "Against Materialism." *Faith and Philosophy* 23, no. 1 (February 1, 2006): 3–32.

Preston, Christopher J. *The Synthetic Age: Outdesigning Evolution, Resurrecting Species, and Reengineering Our World*. Cambridge, MA: MIT Press, 2018.

Priede, Camilla, and Stephen Farrall. "Comparing Results from Different Styles of Cognitive Interviewing: 'Verbal Probing' vs. 'Thinking Aloud.'" *International Journal of Social Research Methodology* 14, no. 4 (July 2011): 271–87.

Quadrato, Giorgia, Tuan Nguyen, Evan Z. Macosko, John L. Sherwood, Sung Min Yang, Daniel R. Berger, Natalie Maria et al. "Cell Diversity and Network Dynamics in Photosensitive Human Brain Organoids." *Nature* 545, no. 7652 (May 2017): 48–53.

Regalado, Antonio. "Chinese Scientists Have Put Human Brain Genes in Monkeys—and Yes, They May Be Smarter." *MIT Technology Review*, April 10, 2019,. https://www.technologyreview.com/2019/04/10/136131/chinese-scientists-have-put-human-brain-genes-in-monkeysand-yes-they-may-be-smarter/.

Renick, Timothy M. "A Cabbit in Sheep's Clothing: Exploring the Sources of Our Moral Disquiet about Cloning." *Annual of the Society of Christian Ethics* 18 (1998): 259–74.

Revah, Omer, Felicity Gore, Kevin W. Kelley, Jimena Andersen, Noriaki Sakai, Xiaoyu Chen, Min-Yin Li et al. "Maturation and Circuit Integration of Transplanted Human Cortical Organoids." *Nature* 610, no. 7931 (October 13, 2022): 319–26.

Rizzo, Mario J., and Douglas Glen Whitman. "Little Brother Is Watching You: New Paternalism on the Slippery Slopes." *Arizona Law Review* 51 (2009): 685–740.

Robert, Jason Scott, and Françoise Baylis. "Crossing Species Boundaries." *American Journal of Bioethics* 3, no. 3 (August 2003): 1–13.

Robinson, Daniel N. "Introduction." In *Scientism: The New Orthodoxy*, edited by Richard N. Williams and Daniel N. Robinson, 1–22. London: Bloomsbury Academic, 2015.

Sample, Ian. "Growing Brains in Labs: Why It's Time for an Ethical Debate." *Guardian*, April 25, 2018.

Sawai, Tsutomu, Hideya Sakaguchi, Elizabeth Thomas, Jun Takahashi, and Misao Fujita. "The Ethics of Cerebral Organoid Research: Being Conscious of Consciousness." *Stem Cell Reports* 13, no. 3 (September 2019): 440–47.

Schermer, Maartje. "The Mind and the Machine. On the Conceptual and Moral Implications of Brain-Machine Interaction." *NanoEthics* 3, no. 3 (December 2009): 217–30.

Scheufele, Dietram A., Nicole M. Krause, Isabelle Freiling, and Dominique Brossard. "What We Know about Effective Public Engagement on CRISPR and Beyond." *Proceedings of the National Academy of Sciences* 118, no. 22 (June 1, 2021), https://doi.org/10.1073/pnas.2004835117.

Seth, Anil K. "Consciousness: The Last 50 Years (and the Next)." *Brain and Neuroscience Advances* 2 (January 2018): 1–6.

Sharp, Lesley A. "Organ Transplantation as a Transformative Experience: Anthropological Insights into the Restructuring of the Self." *Medical Anthropology Quarterly* 9, no. 3 (1995): 357–89.

Shostak, Sara, Jeremy Freese, Bruce G. Link, and Jo C. Phelan. "The Politics of the Gene: Social Status and Beliefs about Genetics for Individual Outcomes." *Social Psychology Quarterly* 72, no. 1 (March 1, 2009): 77–93.

Shultz, Leonard D., Fumihiko Ishikawa, and Dale L. Greiner. "Humanized Mice in Translational Biomedical Research." *Nature Reviews Immunology* 7, no. 2 (February 2007): 118–30.

Simis, Molly J., Haley Madden, Michael A. Cacciatore, and Sara K. Yeo. "The Lure of Rationality: Why Does the Deficit Model Persist in Science Communication?" *Public Understanding of Science* 25, no. 4 (May 1, 2016): 400–14.

Singer, Peter. "The Sanctity of Life: Here Today, Gone Tomorrow." *Foreign Policy*, (September/October 2005): 40–41.

Skloot, Rebecca. *The Immortal Life of Henrietta Lacks*. Broadway Books, 2017.

Slaby, Jan. "The New Science of Morality: A Bibliographic Review." *Hedgehog Review* 15, no. 1 (2013): 46–54.

Smith, Christian, with Michael Emerson, Sally Gallagher, Paul Kennedy, and David Sikkink. *American Evangelicalism: Embattled and Thriving*. Chicago: University of Chicago Press, 2014.

Smits, Martijntje. "Taming Monsters: The Cultural Domestication of New Technology." *Technology in Society* 28, no. 4 (November 2006): 489–504.

Sorell, Tom. *Scientism: Philosophy and the Infatuation with Science*. London: Routledge, 1994.

Stenmark, Mikael. "What Is Scientism?" *Religious Studies* 33, no. 1 (March 1997): 15–32.

Stetka, Bret. "Lab-Grown Mini Brains Can Now Mimic the Neural Activity of a Preterm Infant." *Scientific American*, August 29, 2019, https://www.scientificamerican.com/article/lab-grown-mini-brains-can-now-mimic-the-neural-activity-of-a-preterm-infant/.

Stout, Jeffrey. "Moral Abominations." *Soundings* 66, no. 1 (1983): 5–23.

Streiffer, Robert. "Human/Non-Human Chimeras." In *The Stanford Encylopedia of Philosophy* (Summer 2019 edition), edited by Edward N. Zalta, https://plato.stanford.edu/archives/sum2019/entries/chimeras/, 2019.

Swierstra, Tsjalling, Rinie van Est, and Marianne Boenink. "Taking Care of the Symbolic Order: How Converging Technologies Challenge Our Concepts." *NanoEthics* 3, no. 3 (December 2009): 269–80.

Szasz, Ferenc Morton. *The Divided Mind of Protestant America, 1880–1930*. Tuscaloosa: University of Alabama Press, 1982.

Tan, Tao, Jun Wu, Chenyang Si, Shaoxing Dai, Youyue Zhang, Nianqin Sun, E. Zhang et al. "Chimeric Contribution of Human Extended Pluripotent Stem Cells to Monkey Embryos Ex Vivo." *Cell* 184, no. 8 (April 15, 2021): 2020–32.

Thomas, W. I., and D. S. Thomas. *The Child in America: Behavior Problems and Programs*. New York: Knopf, 1928.

Timmermans, Stefan, and Rene Almeling. "Objectification, Standardization, and Commodification in Health Care: A Conceptual Readjustment." *Social Science & Medicine* 69, no. 1 (July 1, 2009): 21–27.

Tooley, Michael. "Personhood." In *A Companion to Bioethics*, 2nd ed., edited by Helga Kuhse and Peter Singer, 129–39. New York: Wiley-Blackwell, 2009.

Trujillo, Cleber A., Richard Gao, Priscilla D. Negraes, Jing Gu, Justin Buchanan, Sebastian Preissl, Allen Wang et al. "Complex Oscillatory Waves Emerging from Cortical Organoids Model Early Human Brain Network Development." *Cell Stem Cell* 25, no. 4 (October 2019): 558–569.

Valliant, Richard, and Jill A. Dever. *Survey Weights: A Step-by-Step Guide to Calculation.* College Station, TX: Stata Press, 2018.

Van Gulick, Robert. "Consciousness." In *The Stanford Encyclopedia of Philosophy*, edited by Edward N. Zalta and Uri Nodelman, Spring 2018. Metaphysics Research Lab, Philosophy Department, Stanford University, 2018, https://plato.stanford.edu/archives/spr2018/entries/consciousness/.

Waldby, Catherine, Marsha Rosengarten, Carla Treloar, and Suzanne Fraser. "Blood and Bioidentity: Ideas about Self, Boundaries and Risk among Blood Donors and People Living with Hepatitis C." *Social Science & Medicine* 59, no. 7 (October 1, 2004): 1461–71.

Walsh, Nicole C., Laurie L. Kenney, Sonal Jangalwe, Ken-Edwin Aryee, Dale L. Greiner, Michael A. Brehm, and Leonard D. Shultz. "Humanized Mouse Models of Clinical Disease." *Annual Review of Pathology: Mechanisms of Disease* 12, no. 1 (2017): 187–215.

Walton, Douglas. *The Slippery Slope Argument.* New York: Oxford University Press, 1992.

Waters, Brent. "Whose Salvation? Which Eschatology? Transhumanism and Christianity as Contending Salvific Religions." In *Transhumanism and Transcendence*, edited by Ronald Cole-Turner, 163–175. Washington, DC: Georgetown University Press, 2011.

Wehner, Peter. "The Evangelical Church Is Breaking Apart." *Atlantic*, October 24, 2021. https://www.theatlantic.com/ideas/archive/2021/10/evangelical-trump-christians-politics/620469/.

West, John. *Darwin Day in America.* Wilmington, Delaware: ISI Books, 2007.

Williams, Richard. "Generalized Ordered Logit/Partial Proportional Odds Models for Ordinal Dependent Variables." *Stata Journal* 6, no. 1 (February 1, 2006): 58–82.

Williams, Richard. "Understanding and Interpreting Generalized Ordered Logit Models." *Journal of Mathematical Sociology* 40, no. 1 (January 2, 2016): 7–20.

Wolkomir, Michelle, Michael Futreal, Eric Woodrum, and Thomas Hoban. "Denominational Subcultures of Environmentalism." *Review of Religious Research* 38, no. 4 (1997): 325–43.

Wolkomir, Michelle, Michael Futreal, Eric Woodrum, and Thomas Hoban. "Substantive Religious Belief and Environmentalism." *Social Science Quarterly* 78, no. 1 (1997): 96–108.

Wolstenholme, G. E. W., ed. *Man and His Future.* New York: John Wiley, 1963.

Woodberry, R. D., J. Z. Park, L. A. Kellstedt, M. D. Regnerus, and B. Steensland. "The Measure of American Religious Traditions: Theoretical and Measurement Considerations." *Social Forces* 91, no. 1 (September 1, 2012): 65–73.

Worley, Sara. "Materialism versus Dualism." In *The Encyclopedia of Clinical Psychology*, edited by Robin L. Cautin and Scott O. Lilienfeld, 1–8. New York: John Wiley, 2015.

Wuthnow, Robert. *Producing the Sacred: An Essay on Public Religion.* Champaign: University of Illinois Press, 1994.

Wuthnow, Robert. *The Restructuring of American Religion.* Princeton, NJ: Princeton University Press, 1988.

Wuthnow, Robert, and John H. Evans. *The Quiet Hand of God: Faith-Based Activism and the Public Role of Mainline Protestantism.* Berkeley: University of California Press, 2002.

Yap, Kok Hooi, Ralph Murphy, Mohan Devbhandari, and Rajamiyer Venkateswaran. "Aortic Valve Replacement: Is Porcine or Bovine Valve Better?" *Interactive CardioVascular and Thoracic Surgery* 16, no. 3 (March 1, 2013): 361–73.

Zerubavel, Eviatar. *Social Mindscapes: An Invitation to Cognitive Sociology.* Cambridge, MA: Harvard University Press, 1997.

Zerubavel, Eviatar. *The Fine Line: Making Distinctions in Everyday Life.* New York: Free Press, 1991.

Zimmer, Carl. "Human Brain Cells Grow in Rats, and Feel What the Rats Feel." *The New York Times*, October 12, 2022.

Zimmer, Carl. "Organoids Are Not Brains. How Are They Making Brain Waves?," *The New York Times*, August 29, 2019.

Zuckerman, Phil, Luke W. Galen, and Frank L. Pasquale. *The Nonreligious: Understanding Secular People and Societies.* New York: Oxford University Press, 2016.

Index

For the benefit of digital users, indexed terms that span two pages (e.g., 52–53) may, on occasion, appear on only one of those pages.

human-animal chimeras, 1–4, 29
human behaviors barrier, 158–59
human brain organoids (HBOs).
     See also public's views of HBOs
     and NCs
   assembloids and, 5
   caution regarding experiments on,
      10–12, 20
   consciousness and, 5, 19–21
   definition of, 5
   demography and support
      for, 96–98
   developmental experiments on, 5–7
   as having shared thoughts with
      donor, 89–92, 93–94
   humanization of, 19–21
   impact of religious tradition on
      views of, 114–16
   popular portrayal of, 6, 9, 21
   public-policy bioethical debate
      and, 10–14
   survey responses on, 60–63
   vignette on, 55–59, 60–63
humanization
   ephemeral connections to humans
      and, 80–82, 84–85
   HBOs and, 19–21
   NCs and, 19–21, 82, 84–85
   what is to be done and, 134, 137–
      38, 141
human nature. See anthropologies
human-to-animal barrier, 155–56

image of God, 14–16, 116–17
Immortal Life of Henrietta Lacks, The
   (Skloot), 44
impact of religious tradition on
   views of HBOs, 114–16
industrial approach and synthesized
   world, 100–1
interviewees' understanding of
   ephemeral connections to
   humans, 87–88

interviewees' understanding
   of shared thoughts with
   HBO, 90–92

Judaism
   anthropology in, 14, 15–16
   Conservative Judaism, 15–16
   dehumanization and, 133
   dominion beliefs and, 117–18
   foundational distinction between
      humans and animals in, 29–
      30, 32–33
   image of God in, 15–16
   Orthodox Judaism, 27
   science's relation to theology in, 112
   survey responses and, 111

Kass, Leon, 137, 138–39
Keen, Sam, 133
Klitzman, Robert, 2
Knoppers, Bartha Maria, 179–80n.10
Koch, Christof, 21

Lacks, Henrietta, 43, 44–45
Leopold, Aldo, 98–99
levels of scientific knowledge, 123–24
Lewis, C. S., 133
Loike, John D., 27, 180n.11
looking forward. See what is to
   be done

manipulation checks on vignette
   experiments, 171–72
Matrix, The (film), 2–3
Mayr, Ernst, 18
medical research barrier, 159–60
methodology
   attention and, 164–65
   creating religion variables from
      the survey and, 167–70
   manipulation checks on vignette
      experiments and, 171–72
   overview of, 163–72